Chemical Reaction Engineering

Reviews

Chemical Reaction Engineering Reviews

Hugh M. Hulburt, EDITOR,
Northwestern University

The Third International Symposium on Chemical Reaction Engineering co-sponsored by the American Chemical Society, the American Institute of Chemical Engineers, the Canadian Society for Chemical Engineering, and the European Federation of Chemical Engineering, held at Northwestern University, Evanston, Ill., Aug. 27–29, 1974.

ADVANCES IN CHEMISTRY SERIES **148**

AMERICAN CHEMICAL SOCIETY

WASHINGTON, D. C. 1975

Library of Congress CIP Data

International Symposium on Chemical Reaction Engineering, 3d, Northwestern University, 1974.
 Chemical reaction engineering reviews.
 (Advances in chemistry series; 148)

The second and final vol. of the proceedings of this conference.

Includes bibliographical references and index.

1. Chemical engineering—Congresses. 2. Chemical reactions—Congresses.
 I. Hulburt, Hugh M. II. American Chemical Society. III. Title. IV. Series.
 QD1.A355 no. 148 [TP5] 540'.8s [660.2'9'9] 75-42026
 ISBN 0-8412-0272-9   ADCSAJ 148 1-238 (1975)

Copyright © 1975

American Chemical Society

All Rights Reserved

PRINTED IN THE UNITED STATES OF AMERICA

Advances in Chemistry Series
Robert F. Gould, *Editor*

Advisory Board

Kenneth B. Bischoff

Edith M. Flanigen

Jesse C. H. Hwa

Phillip C. Kearney

Egon Matijević

Nina I. McClelland

Thomas J. Murphy

John B. Pfeiffer

Joseph V. Rodricks

FOREWORD

ADVANCES IN CHEMISTRY SERIES was founded in 1949 by the American Chemical Society as an outlet for symposia and collections of data in special areas of topical interest that could not be accommodated in the Society's journals. It provides a medium for symposia that would otherwise be fragmented, their papers distributed among several journals or not published at all. Papers are refereed critically according to ACS editorial standards and receive the careful attention and processing characteristic of ACS publications. Papers published in ADVANCES IN CHEMISTRY SERIES are original contributions not published elsewhere in whole or major part and include reports of research as well as reviews since symposia may embrace both types of presentation.

CONTENTS

Preface ix

1. The Catalytic Muffler 1
   James Wei

2. Kinetic Models in Heterogeneous Catalysis 26
   Sol W. Weller

3. Contacting Effectiveness in Trickle Bed Reactors 50
   Charles N. Satterfield

4. Physical Processes in Chemical Reactor Engineering 75
   E. Wicke

5. Industrial Process Models—State of the Art 98
   Vern W. Weekman, Jr.

6. The Role of Chemical Reaction Engineering in Coal Gasification ... 132
   Herman F. Feldmann

7. Multiplicity, Stability, and Sensitivity of States in Chemically Reacting Systems—A Review 156
   Roger A. Schmitz

8. Oxidation Reaction Engineering 212
   A. Cappelli

Index 233

PREFACE

The present volume completes the proceedings of the Third International Symposium on Chemical Reaction Engineering, held at Northwestern University, Evanston, Ill., August 27-29, 1974. It comprises eight reviews which were invited for the symposium and cover its principal areas of emphasis. Two of the reviews cover specific areas of current interest in chemical processing: automotive exhaust catalysis and the production of clean fuels from coal. Six summarize the state of the art in particular functions involved in chemical reaction engineering. The authors have had early access to the contributed papers, now published as "Chemical Reaction Engineering—II," ADVANCES IN CHEMISTRY SERIES No. 133 (American Chemical Society, Washington, D. C., 1974) and in many cases have built upon and amplified them in assessing the current status of work in the field.

We trust that, as in previous symposia, these reviews will afford a bench mark for assessing past achievements and future progress.

The Editor and Chairman of the Symposium owes a special debt of appreciation to the staff of the Books Department of the American Chemical Society for their patience, skill, and energy in producing the two volumes of papers and reviews efficiently, professionally, and on time.

Northwestern University
Evanston, Ill.
August 1975

HUGH M. HULBURT

Organizing Committee

for the

Third International Symposium on

Chemical Reaction Engineering

H. M. Hulburt, *Chairman*
J. B. Butt, *Co-Chairman*

D. S. Hacker, *Local Arrangements and Events*
W. F. Stevens, *Session Staff and Facilities*
G. M. Brown, *Treasurer*

Members: K. B. Bischoff (AIChE)
D. Thoenes (EFChE)
V. W. Weekman, Jr. (ACS)
J. M. H. Fortuin (EFChE)
I. G. Dalla Lana (CSChE)

The Catalytic Muffler

JAMES WEI

Department of Chemical Engineering, University of Delaware, Newark, Del. 19711

> *The catalytic muffler represents a delightful union of society needs and research opportunity. This new technology is the only effective solution to automotive pollution today, and cannot be effectively challenged by alternative solutions through engine changes till the 1980's. This first generation catalytic muffler has not been optimized in design through many years of experience and is capable of tremendous improvements. Pioneering research is particularly needed to further our understanding in (a) maintenance-free design and fail-safe reliability, (b) transport and modeling of shallow pellet beds and monoliths, (c) optimum design of catalysts and reactors for negative order kinetics, and (d) transient behavior of reactors.*

In the fall of 1974 millions of cars that went on sale were equipped with oxidizing catalytic converters to reduce the emissions of carbon monoxide and hydrocarbons (1). A reducing catalytic converter for NO_x may follow in a few years. The terms "catalytic muffler" and "catalytic converter" are sometimes used interchangeably, which are shortened versions for the more comprehensive term "Kraftfahrzeugabgasentgiftungskatalysator." However, the catalytic muffler does not muffle engine noise and is located much closer to the engine exhaust manifold than a regular muffler.

There are recent comprehensive reviews of the catalytic muffler, covering background, present technology, thermodynamics and kinetics, physical transport processes, and durability (2, 3). The purpose of the present review is to concentrate on the reaction engineering aspects—a few past achievements and a great many unsolved problems.

The Clean Air Act as amended in 1970 is the driving force behind a new chemical reaction technology and the debut of the catalytic converter as an item for mass consumption (4, 5). The conventional reciprocating spark-ignition gasoline engine cannot meet the relaxed federal standards for 1975 model automobiles in California without the use of the oxidizing catalytic converter. In fact, the catalytic converter will lead to more leeway in engine design, which means up to 20% improvement in vehicle mileage per gallon of gasoline and in driveability (suppression of stalling, hesitation, backfire, and other engine malfunctions). Therefore, the catalytic converter was installed on most 1975 model cars to meet the more relaxed federal standards for the other 49 states.

The debut of the catalytic converter is marred by two controversies. Gasoline normally contains 0.03% sulfur by weight, which contributes less than 1% of man-made sources of sulfur in the air. This sulfur is emitted as sulfur dioxide into the atmosphere and is further oxidized to sulfur trioxide by the catalytic action of metal ions in aerosols and by sunlight (6). The resulting sulfuric acid and sulfates are considered greater health hazards than sulfur dioxide. The present ambient air standard for sulfur dioxide is 365 $\mu g/m^3$, but the proposed standard for sulfates is 10 $\mu g/m^3$. The catalytic converter oxidizes 10–30% of the input SO_2 to SO_3, which is only a small additional burden when properly dispersed (7), but it may give rise to high local concentrations of sulfates along heavy traffic lanes. Much more definitive investigations are needed to determine whether the benefits of reducing CO emission from 30 to 3.4 g/mile, and hydrocarbons from 3.4 to 0.41 g/mile, is greater than the harm of increasing sulfate emission from 0.001 to approximately 0.03 g/mile while recognizing that much of the 0.16 g/mile emission of SO_2 would oxidize to sulfates in the atmosphere later. The Environmental Protection Agency is contemplating a controversial vehicle emission standard of sulfates at no more than 0.001 to 0.01 g/mile, which could lead to the demise of the catalytic converter. This proposal is challenged by the Federal Energy Administration and the State of California. Existing technology can save the catalytic converter by reducing the sulfur content of gasoline, by improving the absorption of sulfates in the converter, and by reducing the excess air in the converter to suppress SO_3 formation, with a cost increase. Another source of concern is the outside skin temperature of the catalytic converters, which may reach 900°F after a hill climb at full throttle. This converter temperature is only somewhat higher than a conventional muffler at the same location, and 400 degrees lower than that reached by the exhaust manifold of the engine, but the converter is located closer to the ground and may cause fire and explosion over tall grass and combustibles. This has led to the banning of catalytic converters by some oil refineries and chemical plants.

An engine change is the other approach to meet the Federal standards on carbon monoxide and hydrocarbons (8). It is far more difficult to meet the standards on NO_x. The diesel engine emits smoke and odor and is 50% larger and heavier than the gasoline engine. The rotary engine is already in mass production for the subcompact Mazda, which requires a thermal afterburner and provides a disappointing gasoline mileage. The stratified charge engine of Honda Motors has shown great promise but needs proof of durability. There are many other interesting new engines in the research and early development phase, including the steam engine, the Stirling engine, the Rankin engine, the gas turbine, the fuel cell, and the storage battery electric engines. All have strong points, and all have great weaknesses that must be overcome. Experts in manufacturing engineering and in customer service, within and outside of the automotive industry, believe that these engines cannot capture a large percentage of the total market before the 1980's. In the remainder of this decade, the choice is among the conventional engine with the catalytic converter, an enforced reduction of permissible vehicle-mileage in urban areas, and a further reduction in the standards of ambient air quality. For the next decade, the best solution is the goal of a vigorous competition between the investigators of new engines and the researchers of the catalytic converter.

Platinum and palladium are the active ingredients used in the first generation of catalytic converters. In one design, these noble metals are deposited on 1/16- to 1/8-inch pellets of high surface area alumina and placed in a very shallow bed. In another design, the noble metals are deposited on a thin "wash coat" of alumina, which adheres to the walls of ceramic honeycomb monoliths. Since space is at a premium in an automobile, the reactor volume should be kept to a minimum so that it will be compact enough to fit inside the already crammed engine hood, or it should be flat enough to fit under the front passenger's seat.

The stated goal of the law is to minimize emissions from each new car over 50,000 miles of use, subject to cost and reliability constraints. Since urban pollution is the main target, the authorized test cycle simulates an urban car starting in the morning with a cold engine and going through stop-and-go traffic. The certification procedure specifies the maximum permissible grams emission per mile traveled, regardless of automobile weight or engine size.

$$W = \int mF(t)c(t) / \int V(t)dt \text{ (g/mile)} \tag{1}$$

where m is the molecular weight of the pollutant in g/mole, F is flow rate of exhaust gas in moles/min, c is pollutant concentration in mole fraction, and V is vehicle speed in miles/min. The currently official CVS-CH procedure (Federal Cycle) calls for a cycle from a cold start, with a rigidly specified schedule of vehicle speed as a function of time, covering 7.5 miles in 22.9 minutes. This is followed by a shutdown for 10 minutes and a re-start and repeat of the first 8.4 minutes of the cycle.

The dynamic range of the input variables to the catalytic converter is impressive. The top speed attained in the Federal Cycle is 56.7 mph and can be considerably higher in road use. The inlet gas temperature varies from ambient to 1200°F in the Cycle and may go up to 1800°F in road use. The chemical heat contained in the exhaust gas can be considerable since each mole percent of CO yields 140°F temperature rise upon oxidation. The flow rate of gases vary from 10 SCFM during engine idle to 100 SCFM during rapid acceleration in the Cycle and may go up to 200 SCFM in road use. The gaseous composition can vary from an oxidizing condition with a lean air-to-fuel ratio to a reducing condition with a rich ratio. A secondary air pump is often needed with enough capacity to ensure a net oxidizing atmosphere. The CO concentration can vary from below 0.1% to above 8%. Under these conditions, the gas has a reactor residence time of 10 to 200 msec, and a conversion level of 80–90% is needed.

Many combinations of catalyst formulations and reactor configurations are adequate when they are newly installed. The performance deteriorates with use, mainly because of high temperature and poisons. Instead of today's gasoline with 3.0 g of lead per gallon, special gasoline with less than 0.05 g of lead would be needed to avoid early deactivation. This "lead-free" gasoline requires more refining and an additional cost of about 0.1¢/gal (9) according to an independent study, but oil industry estimates are higher than 1¢/gal. The catalytic converter occasionally fails from physical attrition of the catalyst support and mechanical failure of the auxiliary equipments.

In a few years the dollar sale of automotive catalysts may exceed the combined sale of catalysts to the chemical and petroleum industries (10). The application of reaction engineering principles to this emerging technology is still in its infancy. The optimum design of an industrial reactor evolves after many years of experience. There is virtually no experience with the catalytic muffler, and there is tremendous room for improvement. The catalytic converter also presents four new challenges to the art and science of reaction engineering, which calls for pioneering research:

(a) It is mass produced and placed directly in the hands of the consuming public, who cannot and perhaps have no incentive to provide monitoring and maintenance. A maintenance-free design with fail-safe reliability is needed.

(b) Pressure drop across the catalytic converter must be severely limited, requiring the unfamiliar shallow pellet bed, the ceramic monolith, and the metallic screens. The transport properties and modeling of these reactors need intensive study.

(c) The oxidation of carbon monoxide and hydrocarbons over platinum and palladium exhibits a negative order kinetics. The optimal design of catalysts and reactors for such kinetics should be investigated.

(d) The performance of the catalytic converter is defined over a wide dynamic range of input variables and is dominated by the first two minutes of transience from a cold start of the engine. The transient behavior of converters should be better understood.

The catalytic muffler represents a delightful union of society needs and research opportunity. It deserves the attention of some of the best talents in our profession.

Optimum Policy

The overall goal of the Federal Clean Air Act should be the best tradeoff between the benefits from a decrease in air pollution and the costs to the public. Lawmakers need to have their views broadened by a range of proposed policy alternatives, and they need to know the positive and negative consequences of each alternative. Without the inputs from many well-informed, unbiased, and ably argued position papers representing the technical point of view, the public laws enacted may turn out to be wasteful, obstructive, or counter-productive. The optimal design of a set of public laws, aimed at reducing pollution at minimum cost to the nation, should be a high order of business. Reaction engineers who seldom venture into the calculus of costs and benefits and into the formulation of the optimum policy could perform another important duty as experts and concerned citizens.

Ambient air quality depends on the average emission per car and the total vehicle mileage. It does not seem reasonable to forbid the sale of a vehicle that emits more than 9.0 g/mile of CO and to accept with equal pleasure two vehicles emitting 1.0 and 8.9 g/mile. Such a law will lead to the following distortions: encouraging a design for 8.9 g/mile, with allowances for the variability in manufacturing quality control and in vehicle testing experimental scatter; condemning to a junk yard an automobile that emits 9.1 g/mile; giving no incentive to innovations leading to very low emission cars. Instead, a law that sets a penalty that is a continuous and rapidly increasing function of emission, coupled with a positive reward for very low emission, would induce the manufacturers to produce better cars.

In a remarkable article, Shinnar argued that to obtain the highest air quality, the emission standards must not be set too low (*11*). An emission standard that is too stringent would be self-defeating since it requires an elaborate engine design that is more prone to failure. A converter failure may bring an emission that is 10–20 times the allowed limit. If 5% of the automobiles on the road have failed converters that are not yet detected and corrected, then the ambient air quality is dominated by the failure rate. A more stringent standard would lead to a higher failure rate and dirtier air.

There are also many related areas where reaction engineers could offer their time and talent as concerned citizens. For instance, why should the air quality standard for CO specify that anywhere in the United States an 8-hr average concentration of 9 ppm by volume should not occur more than once a year? Why should the standard of automobile emissions, tailored to cure urban ills, be imposed on rural drivers? Since 15 ppm is good enough for healthy persons, would it not be sensible to relocate heart disease patients away from center city?

The task of curbing automobile pollutants cannot be ended by regulating only the automobile manufacturers. There is no current federal law to require the periodic inspection and maintenance of vehicles in use. In the chemical industry, a catalytic reactor is often a multimillion dollar investment that is designed and constructed with the expenditure of many engineering man-hours. To maximize profit, an industrial reactor is monitored by many measuring and recording instruments, under the watchful eyes of plant engineers ready to make corrective measures. In contrast, the catalytic muffler with its accessories has a retail value of perhaps $150 and is designed and constructed for the daily use of the general public. The continued performance of the catalytic converter may be a matter of indifference to the average motorist, who in any case has neither the instrumentation nor the skill to inspect and to maintain. The catalytic muffler needs to be failure-proof for 3 to 5 years without attention—a feature not shared by any automobile accessory.

It has been proposed that the lack of motivation to repair a malfunctioning catalytic muffler may be augmented by the installation of a device that emits loud noises which cannot be silenced till the catalytic muffler is repaired. This solution is still inadequate since the repair service industry has not yet been organized. Thousands of mechanics must be trained to diagnose and to repair the various types of catalytic mufflers, thousands of diagnostic instruments must be mass produced and distributed, millions of spare parts must be produced and stockpiled in numerous warehouses (*12*). These tasks have not yet begun. Public law is too serious a business to be left to the lawyers alone.

Unusual Catalytic Beds

Because of the concerns for engine performance and fuel economy, the pressure drop across the catalytic muffler must be kept to the minimum. The highest pressure drop occurs on wide-open throttle engine operations when the exhaust gas flow rate and temperature are at the maximum. A pressure drop of 5–8 inches of water is regarded as the upper limit of acceptability. This requirement has led to the development of very shallow packed beds and monoliths.

Shallow Packed Beds. The current design of packed beds have "pancake" aspect ratios, typically 1–2 inches deep with a cross-sectional area of some 100 square inches. Since the pellets have a diameter of 1/16 to 1/8 inch, this means a bed of 10 to 20 layers of particles, which is a radical departure from industrial beds of hundreds or thousands of layers. Pressure drop requirements have also led to the development of a number of radial-flow reactors, where the catalysts are placed in the annulus of two concentric cylinders and where the gas flows in a direction parallel to the radius and transverse to the axis of the cylinders (*see* Figure 1). Since both axial-flow and radial-flow reactors will be considered, the two directions in the reactor will be referred to as the "longitudinal" and the "transverse" directions to avoid confusion.

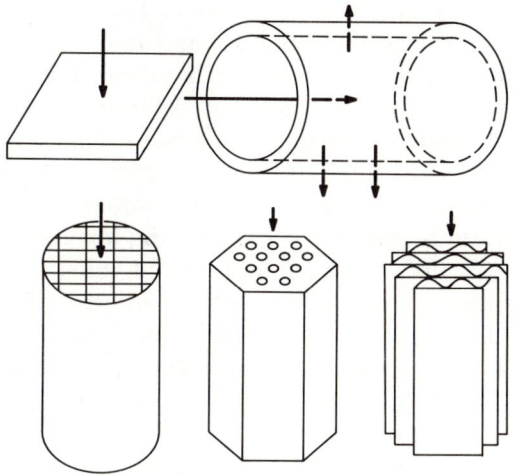

Figure 1. Shallow pellet beds and monoliths

When a packed bed is less than 50 particles deep, a step signal created by a sudden change in inlet concentration spreads as it travels through the bed (*13*). The longitudinal spreading of a temperature signal is considerably stronger (*14, 15*). These modifications to the concentration and temperature profiles have a strong effect on the reactor performance. Dispersion of mass in the transverse direction is unimportant in a catalytic muffler since the side walls are impervious and noncatalytic. Transverse dispersion of heat is insignificant for the short and fat "pancake" reactors which approach adiabatic conditions but is more important for the long and slim "cigars" reactors. Theories and experiments on the modeling of such shallow packed beds under dynamic conditions are not well developed.

Using the terminology of Froment, the minimum model to use is the Pseudo-Homogeneous A. II, where longitudinal dispersions are imposed on a piston flow reactor (16):

$$\begin{cases} D \dfrac{d^2c}{dx^2} - u \dfrac{dc}{dx} - R = 0 \\ \lambda \dfrac{d^2T}{dx^2} - u(\rho c_p) \text{ gas } \dfrac{dT}{dx} + H \cdot R = 0 \end{cases} \quad (2)$$

The gases flow through an exhaust pipe of 2 to 3 inches in diameter in a pulsating flow, with a frequency equal to half of the rpm of the engine, multiplied by the number of cylinders and divided by the number of converters. A rapidly expanding cone connects the exhaust pipe with the catalytic bed of some 100 square inches in cross-sectional area, and a similar contraction cone gathers the exit gas into an exhaust pipe. The exhaust gas flow is turbulent in the exhaust pipe upstream from the catalytic bed, with a Reynolds number of 5000–80,000. The Danckwerts boundary conditions may be inappropriate since longitudinal dispersions of mass and heat occur outside of the reactor in an unknown manner.

Inside the packed bed the Reynolds number of the gaseous flow in the Federal Cycle is from 10 to 200, and the longitudinal Peclet number, ud_p/D, for mass is almost constant at the value of 2 (13). This is very fortunate since one can then use the same fixed Peclet number at all flow rates. On the other hand, the longitudinal Peclet number for heat, $ud_p(\rho c_p)\text{gas}/\lambda$, is approximately constant with a value of 0.3 (15). This sevenfold difference in the values of mass and heat longitudinal dispersion complicates the computations.

The Pseudo-Homogeneous Cascade model is a good deal more amenable to numerical computation.

$$\begin{array}{ll} u(c_{i-1} - c_i) - R = 0 & i = 0 \text{ to } N \\ (\rho c_p)_g u(T_{i-1} - T_i) + R \cdot H = 0 & \end{array} \quad (3)$$

When the number of cells in the Cascade is chosen to be $N = uL/2D$, the distribution of gaseous residence times is very similar between the two models provided that N is greater than 10. Therefore to simulate mass dispersion when the longitudinal Peclet number is 2, N should be chosen to equal L/d_p, or each cell should consist of one layer of catalyst in the bed. On the other hand, to simulate temperature dispersion when the longitudinal Peclet number for heat is 0.3, N should be chosen to equal $L/7d_p$, or each cell should consist of seven layers. For typical beds where L/d_p is about 15, N should be equal to 15 or 2, depending on whether mass dispersion or heat dispersion should be emphasized. With such a small value of N, the analogy between the Dispersed Plug Flow and Cascade model breaks down.

Caught between these two requirements, one might argue that an accurate evaluation of heat dispersion is more important than an accurate evaluate of mass dispersion because of the Arrhenius expression for reaction rate. To be more precise, we can define the relative sensitivity of the reaction rate to a percentage change in temperature compared with a percentage change in concentration as:

$$S = \frac{\partial R}{\partial \ln T} \bigg/ \frac{\partial R}{\partial \ln c}$$

when

$$R = f(c) e^{-Q/R_g T} \qquad (4)$$

$$S = \frac{Q}{R_g T} \bigg/ \frac{d \ln f}{d \ln c}$$

If the reaction rate is a simple n-th order,

$$\frac{d \ln f}{d \ln c} = n$$

so that $S = Q/R_g T \cdot 1/n$ and has a value of about 20 for $n \leq 1$. It is justified to choose N according to the needs for thermal dispersion. The best value for N was determined by a set of transient experiments to be $N = (L/d_p) x (1/3)$, or each cell should consist of three layers of catalysts (17, 18). Experimental values of the longitudinal dispersion in a radial-flow reactor were published by Hlavacek et al. (19, 20).

The Heterogeneous Cascade model is also used for this problem, where the temperature and concentration of the solid and gas phases are allowed to be different. Dispersion is related to the gas to solid transfer coefficients as well as to N.

$$u(T_{i-1}^g - T_i^g) = ha(T_i^g - T_i^s) + H \cdot R \qquad (5)$$

However, the rates of mass and heat transfer between the solid and the gas are not proportional to gaseous flow rates. The j_H and j_D factors vary approximately as -0.5 power of the Reynolds number so that Nu and Sh vary as 0.5 power of Reynolds number, leading to transference units that vary with -0.5 power of Reynolds number.

$$\mathrm{TU}_H = \frac{\mathrm{Nu} \cdot aL}{\mathrm{Re} \cdot \mathrm{Pr}} \qquad \mathrm{TU}_M = \frac{\mathrm{Sh} \cdot aL}{\mathrm{Re} \cdot \mathrm{Sc}} \qquad (6)$$

All the above theoretical models assume uniform flow distributions. In practice, channeling could be a problem in view of the sudden expansion and contraction in gas flow. It could be remedied by judicious use of baffles and deflectors, at the expense of increased pressure drop. Thermal radiation effects are very important for the "pancake" design.

Monolithic Catalytic Beds. A bundle of parallel tubes offers much less resistance to air flow than a random packed bed of spheres since the flow does not have to change direction repeatedly and to split up and rejoin around each sphere. The ceramic monolith is an integral bundle of tubes with a variety of cross-sectional shapes including the circle, the hexagon, the square, the equilateral triangle, and the sinusoid. In one process a mixture of fibers and ceramic material is extruded from a die and then fired in a kiln. In another process the material is formed into alternating sheets of flat and corrugated layers, arranged into a stack, and then fired in a kiln. Since these ceramic products have too little surface area to support and to disperse catalysts, a high surface area alumina is deposited on the tube surface as a "wash coat." A number of open metallic catalyst supports have also been developed in the form of open-mesh and reinforced wire structures and staggered layers of metallic screens or saddles. The pressure drop through the monoliths and metallic screens tends to be lower so that these catalytic beds may have "cigar" aspect ratios, typically 3–5 inches in diameter and 2–9 inches in length.

A typical monolith used in catalytic mufflers has a channel diameter of 0.05 inch, a wall thickness of 0.01 inch, and a washcoat thickness of 0.001 inch where the active precious metals are deposited. It is contained in a stainless steel cylinder 5 inches in diameter and 5 inches long, connected to the 2½-inch exhaust pipes by short cones. Gas flow rates generated during the Federal Cycle would give Reynolds number ranging from 20 to 400, which leads to a streamline flow at all times except perhaps at the front opening of the channel where the gas may still be turbulent. A great deal of channeling takes place in the monolith, induced by the rapidly expanding and contracting cones. The flow rate is many times greater at the center of the axis, causing faster warmup from a cold start, faster aging, and lower conversion efficiency. This channeling can be reduced by using flow deflectors upstream in the expansion cone, with a simultaneous increase in pressure drop. A number of devices have been tried and reported in the literature (*21, 22*).

The rates of mass and heat transfer from gases in streamline flow to the walls have been investigated by many people beginning with Graetz and Nusselt, often in connection with the design and operation of compact heat exchangers (*23*). The rate of heat transfer is important during the warmup, and also during destructive overheating of the monolith at high temperature operations. The early theoretical solutions consider circular channels with fully developed velocity profiles, neglecting longitudinal dispersion in the gas phase and radial conduction in the solid phase.

$$u(\rho c_p)_g \frac{\partial T}{\partial x} = k \left[\frac{\partial^2 T}{\partial r^2} + \frac{1}{r} \frac{\partial T}{\partial r} \right] \quad (7)$$

$$u(r) = 2\bar{u}(1 - r^2/r_o^2)$$

The solution is

$$T(r,x) = \Sigma c_i \psi_i(r/r_o) \exp\left(-\beta_i^2 \frac{\alpha x}{2ur_o \cdot r_o}\right) \quad (8)$$

where the eigenvalues are $\beta_0 = 2.705$, $\beta_1 = 6.667$, $\beta_2 = 10.67$, etc. and the eigenfunction ψ_i has i zeros according to the Sturmian oscillation theorem. When x is sufficiently large, all the terms in the series become insignificant compared with the first term, and the radial temperature profile reaches constant shape. It is customary to define a heat transfer coefficient, h, in terms of the mixing cup average temperature of the gas and to compute the Nusselt number Nu = hd/λ. At the entrance of the channel, the higher terms contribute to a very large value of Nu; when the value of $x/d \cdot \text{Re} \cdot \text{Pr} \geq 0.05$, Nu reaches the asymptotic value governed by the eigenfunction ψ_0 alone. These solutions were extended to channels with a plug flow entrance and a developing velocity profile. The two boundary conditions often used are constant wall temperature and constant wall flux. The value of Nu for developing flow and constant heat flux is greater than Nu for developing flow and constant wall temperature, which is in turn greater than Nu for developed flow and constant wall temperature. The asymptotic Nusselt numbers are independent of the channel diameter or the Reynolds number. When the Reynolds number is increased, the only effect is to lengthen the entrance region of enhanced Nusselt number. In the catalytic muffler, the length of the enhanced region is at most 0.7 inch vs. the channel length of 3–5 inches.

The computation of the Graetz-Nusselt problem was extended to include a large variety of channel geometries, usually requiring numerical techniques. The asymptotic Nusselt numbers were the goals, and velocity profiles were sometimes obtained. The diameter, d, used in the definition of the Nusselt number becomes the hydraulic diameter, or four times channel cross-sectional area divided by the wetted perimeter. These solutions are summarized by Kays and London (24) and by Shah and London (25). They range from 2.5 for sinusoidal channels to 7.5 for parallel plates. The fully developed velocity field in sinusoids was solved by Sherony and Solbrig (26). Schoenherr *et al.* have shown that triangular and sinusoidal channels have stagnant corners where the local heat transfer coefficient almost drops to zero (27).

There is little experimental confirmation of these in the literature. For the catalytic muffler, the conditions of these calculations are not fulfilled: there is turbulence at the entrance of the channel, and the wall is neither at constant temperature nor possesses constant heat flux. Heck et al. have a paper in this symposium, measuring the experimental temperature profiles of a monolith under warmup conditions (28). They found that the average Nusselt number is a mildly increasing function of the Reynolds number and falls between the asymptotic values for constant wall temperature and for constant wall flux. The radial temperature gradient is negligible. The entrance region of developing thermal boundary layer is longer than expected, possibly because of entrance turbulence. Koch studied mass transport in monoliths of various lengths by ethylene oxidation (29). He found good agreement between his data and the fully developed laminar flow theory when the temperature is sufficiently high and mass transport is rate limiting.

The monolith under present muffler design is very seldom under mass transfer limiting conditions, except for turnpike driving conditions with very high speed gas flow and fully warmed catalysts. Assuming an infinitely fast kinetics, the exit concentration from a monolith is given by

$$c/c_o = e^{-TU} = \exp\left[-\frac{\text{Sh} \cdot aL}{\text{ReSc}}\right] \qquad (9)$$

When 90% conversion is desired, the value needed for TU is 2.3. The value of the Sherwood number may be assumed equal to the Nusselt number. Under reaction conditions, the Schmidt number is 0.23 for hydrogen, 0.81 for CO, and 1.60 for benzene. A typical monolith with 0.05-inch channel diameter and 5-inch length has a value of 260 for aL. Thus even with the most unfavorable channel geometry and Sh = 2.5, we would be mass transport limited only when Re = 180, at the upper end reached in the Federal Cycle. However, good heat transfer is needed in the warmup period from a cold start.

The modeling of simultaneous mass and heat transport and chemical reaction on catalytic walls may be approximated by the Heterogeneous B.I model of Froment. The problem is quite complex and can be solved only by numerical techniques even if the Nusselt number is assumed constant throughout the channel length. When the radial gradient in the gas phase is also taken into account, we arrive at a model that is not in the classification scheme of Froment. Under isothermal conditions, this problem was tackled by Katz, by Solomon and Hudson, and by Lupa and Dranoff, using eigenfunction expansion (30, 31, 32). These methods are not widely used since the catalytic muffler channels are not iso- Young and Finlayson solved this problem by using numerical integration, and by orthogonal collocation methods (33). They discovered that a spot may develop on the wall to ignite the reaction, and to rejuvenate the Nusselt number for a short length. If the wall flux suddenly switches from negative to positive, the Nusselt number can decline rapidly to negative infinity and return to the positive asymptotic value by way of positive infinity. This curious phenomenon arises when the average temperature of the gas is higher than wall temperature, which is in turn higher than the gas temperature immediately adjacent to the wall; this gives rise to a positive value for gaseous temperature gradient at the wall but a negative value for T_S-T_g. Experimental investigations are needed to evaluate the practical significance of these phenomena.

An approximate equivalence between the Heterogeneous Dispersion B.II model and the Pseudo-Homogeneous Dispersion A.II model was given by Vortmeyer and Schaefer (34). They demonstrated that when $\partial^2/\partial x^2 (T_S\text{-}T_g) = 0$, model B.II is equivalent to model A.II when one assigns

$$\lambda_A = \lambda_B + u^2(\rho c_p)_g{}^2/ha \qquad (10)$$

This assumption is valid under mild conditions but breaks down under light-off conditions. After transforming the gas–solid heat transfer coefficient into a longitudinal dispersion coefficient, one can take one more step and transform this A.II model into a Cascade model. When $\lambda_B = 0$, let

$$\text{Pe}_H = \frac{ud(\rho c_p)}{\lambda_A} = \frac{dha}{(u\rho c_p)} = \text{Nu}\,\frac{a\alpha}{u}$$

and,

$$N = \frac{\text{Pe}_H}{2}\frac{L}{d} = \frac{d \cdot ha}{2u\rho c_p} \cdot \frac{L}{d}$$

Unfortunately, h is practically independent of Reynolds number instead of being proportional to it. Thus, under low Reynolds number the monolith may be considered to be 60 cells in series, but under high Reynolds number the monolith must be considered to be six cells in series, which makes computation difficult.

The best channel cross-sectional shape, the best channel diameter, and the best channel length are the subjects of several investigations. Sufficient geometric surface area must be provided to disperse the alumina washcoat and platinum. Beyond that point, more surface area would mean a heavier monolith and is not desirable. A small channel diameter would simultaneously increase heat transfer coefficient h, and surface to volume ratio a since the Nusselt number is practically independent of channel diameter; it would also lead to a much higher pressure drop. The best diameter is the result of a compromise. The optimum channel shape is more controversial since there is no generally accepted set of variables to be held constant during a comparison. Circular channels are not seriously considered since they involve thick and heavy walls. For fast warmup, a set of infinitely wide parallel plates would be the best since it would give the highest transfer coefficient to the surface for the same gas flow rate and pressure drop. This configuration must be ruled out for its structural weaknesses. The next best shape is an elongated rectangle, which is still difficult to make and to protect. Hegedus made a comparison for various practical shapes, based on the same hydraulic diameter and average gaseous velocity (*35*). He showed that from the point of view of lower surface area requirement and lower pressure drop, the circle and hexagon are superior to the square, which is superior to the triangle and sinusoid. His results are given in Table I.

Table I. Comparison of Monolith Geometries with the Same Hydraulic Radius[a]

	Circle	*Hexagon*	*Square*	*Triangle*	*Sinusoid*
Volume of wall needed, in.^{3}	10.2	10.2	12.9	15.7	15.7
Pressure drop, inch water	4.1	3.9	4.5	5.1	5.0

[a] Surface needed for 99% conversion under mass transfer limited conditions. All geometries have the same hydraulic radius of 0.062 cm, mass flow rate of 0.75 g/cm^2 sec, wall thickness of 0.0305 cm, and temperature at 600°C. Hegedus defines the hydraulic radius as one-half of hydraulic diameter.

Table II. Comparison of Monolith Geometries with the Same Cross-Sectional Area[a]

	Circle	*Hexagon*	*Square*	*Triangle*	*Sinusoid*
Length, in.	4.4	4.4	4.1	4.1	3.7–4.0
Wall volume, in.^{3}	12.8	9.2	9.8	10.4	11.6
ΔP, inch H$_2$O	4.5	4.7	5.0	5.7	5.5

[a] Length and surface area needed for 99% conversion under mass transfer limited conditions, with open area of 0.02 cm^2.

On the other hand, Johnson and Chang made a comparison based on the assumption of equal channel open cross-sectional area (*36*). A sinusoid, in comparison with a hexagon with the same cross-sectional area, has a smaller hydraulic diameter and larger pressure drop but can be made more compact. Their results are shown in Table II.

These differences are significant for the requirement of fast warmup. During the manufacture of the ceramic monolith, slumping of the wet material will distort the exact geometric shapes; however, thick coatings of washcoats will round out the sharp corners and improve the performance.

Negative Order Kinetics

For most reactions under isothermal conditions, the reaction rate is an increasing function of the concentration of each reactant consumed. Most of our reaction engineering literature is rooted on the Guldberg-Waage mass action law, on reactions of positive fractional order, on rates involving a quotients of polynomials, all obeying the relation $\partial R/\partial C \geqq 0$ where C is the concentration of a reactant consumed. Many of the reaction engineering rules that have been worked out and applied widely are based on the positive order kinetics, such as the economy of the piston flow reactor over the stirred tank reactor in reactor volume requirements, the superiority of the once-through reactor over the recycle reactor, and the decline of the effectiveness of porous catalysts when isothermal diffusion effects are encountered. These rules are reversed when one encounters negative order kinetics.

Negative order kinetics are rare but are industrially important. Autocatalytic and free radical reactions are often cited but are improper examples since a decline in reactant concentration is not always associated with an increase in catalytic or radical concentrations. Hydrogenolysis, or the splitting of a saturated hydrocarbon into two saturated hydrocarbons by hydrogen, exhibits negative order with respect to hydrogen concentrations over many supported metals.

$$AB + H_2 = AH + BH$$

Morikawa, Benedict, and Taylor have found the reaction to be -1.2 to -2.5 order dependent on concentration (*37*). Sinfelt gave the reaction orders of ethane hydrogenolysis over the transition metals as -0.8 to -2.5 (*38*).

The kinetics of CO oxidation over platinum and palladium have been investigated by numerous scientists, beginning with the pioneering work of Irving Langmuir (*39*). He showed that the kinetics can be represented by the expression

$$\text{rate} = k(O_2)/(CO) \tag{11}$$

He speculated that the rate-controlling mechanism is the combination of the adsorbed oxygen atoms and adsorbed CO. The platinum surface is intensively covered with adsorbed CO so that an increase of CO concentration would decrease the adsorbed concentration of oxygen and the rate of reaction. This kinetic phenomenon has been confirmed many times by various investigators (*40, 41, 42, 43, 44*). Baddour and Cochran studied CO oxidation over palladium and platinum with simultaneous infrared measurements (*45*). They discovered that the kinetics are divided into two regimes: a "normal regime" where the CO concentration is above a critical value, where the kinetics has a negative dependence on CO concentration, and where there is significant infrared absorption at 2100 cm^{-1}; and a "low surface coverage regime" where the CO concentration is below a critical value, where the kinetics is first order to CO concentration, and where the infrared absorption peak is absent.

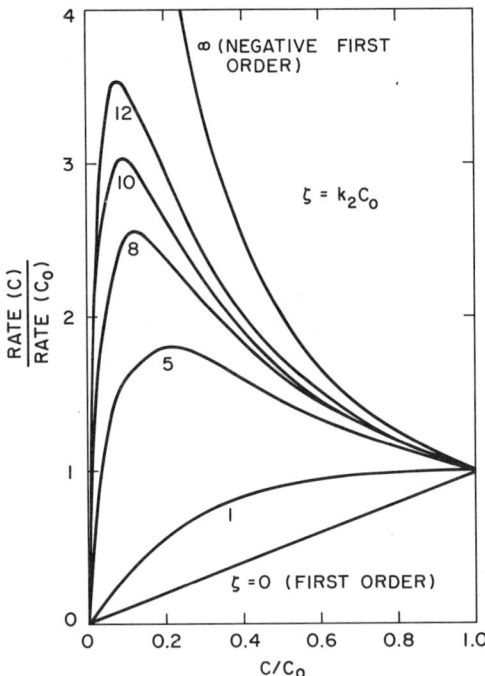

Figure 2. The bimolecular Langmuir kinetics

A careful experiment and quantitave analysis of the kinetic data was performed by Voltz *et al.* (*46*). Their experiment was done over impregnated platinum catalysts on pellets, with temperatures and gaseous compositions simulating exhaust gases. To minimize diffusion effects, most of the platinum was deposited in the outer portions of the particle. They found that their CO oxidation data can be fitted by

$$\text{rate} = \frac{k_1 (CO)(O_2)}{[1 + k_2(CO)]^2} \tag{12}$$

where (CO) and (O_2) are in mole %-atm and
$k_1 = 1.83 \times 10^{12} \exp(-22{,}500/T)$, $(\sec)^{-1}(O_2)^{-1}$
$k_2 = 6.55 \times 10^{-1} \exp(+1{,}730/T)$, $(CO)^{-1}$
T = temperature, degrees Rankin.

The denominator term in Equation 12 contains contributions from NO and hydrocarbons when they are present. The propylene oxidation rate has the same form as Equation 12. The form of these kinetics is given in Figure 2 which is strikingly different from the form of a first-order kinetics. Here, C is the concentration of CO. The maximum rate is attained at $C = 1/k_2$.

Many peculiar properties result from the bimolar Langmuir kinetics of Equation 12. For instance, the concentration profile of a first-order kinetics in an isothermal piston flow reactor follows an exponential curve shown in Figure 3. This convex function indicates that any backmixing would harm the performance so that a constant stirred tank reactor or a recycle reactor of the same volume would be far less efficient than a piston flow reactor. However for a bimolecular Langmuir kinetics in the same reactor, the concentration profile is given by integration of Equation 12 to yield:

$$k_1(O_2)\tau = -\ln y + 2\zeta(1 - y) + \zeta^2(1 - y^2)/2 \tag{13}$$

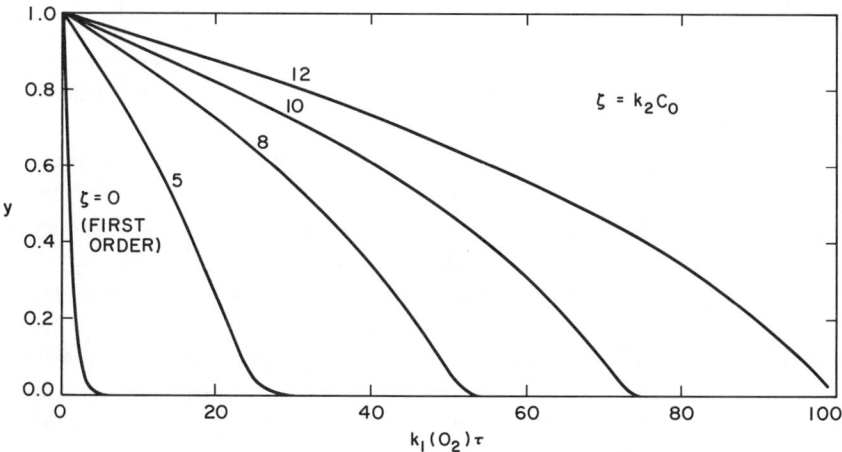

Figure 3. Conversion in piston flow reactor

where τ is the residence time, y is the ratio $(CO)/(CO)_{inlet}$, and ζ is $k_2 C_{inlet}$.

The concentration profiles given in Figure 3 are concave so that backmixing would be beneficial (3). When the bimolecular Langmuir kinetics occur in an isothermal CSTR, the reactor performance is described by

$$k_1(O_2)\tau = (1-y)(1+\zeta y)^2/y \qquad (14)$$

Above a critical value of $\zeta = 8$, there will be three steady-state solutions for each value of $k_1(O_2)\tau$. Thus in an experiment where the value of $k_1(O_2)\tau$ is varied by changing the flow rate of gases or by changing the flow rate of gases or by changing the reactor temperature, the degree of conversion would take sudden leaps and would exhibit hysteresis (see Figure 4). For the recycle reactor with a recycle ratio of R, the reactor equation is:

$$k_1(O_2)\tau = (1+R)\{\ln z + 2\zeta y(z-1) + \zeta^2 y^2(z^2-1)/2\} \qquad (15)$$

where $z = (1+Ry)/(1+R)y$

Similar instabilities can be observed in Figure 5.

Thus, the CSTR and the recycle reactors are superior in performance to the piston flow reactor when the kinetics are of the bimolecular Langmuir type. Great care must be taken to avoid the instabilities which can arise during the acceleration of an automobile when the residence time can rise by a factor of 20, when the CO inlet concentration may vary from 8 to 0.1%, and when the temperature may vary from 900° to 1800°F (3).

An even more interesting phenomenon arises when the bimolecular Langmuir kinetics take place with simultaneous diffusion across a porous catalyst. The classical Thiele analysis showed that a drop of concentration in the interior of a catalyst would lead to a decrease in overall reaction rate—as long as there is a positive dependence of local reaction rate on local concentration. For the bimolecular Langmuir kinetics the same concentration drop in the interior would lead to an increase in overall reaction rate (47, 48). Wei and Becker (49) showed that even in an isothermal reactor the overall effectiveness factor can be greater than 1 and that multiple steady-state solutions can be expected when the value of ζ exceeds 8. This means that for a given quantity of platinum it is better to spread it over a thicker support layer than to concentrate it on a narrow layer near the pellet surface. An even deeper placement of platinum is advantageous in avoiding lead and phosphorus deposition on the catalyst pellets, which tend to concentrate on the surface (50).

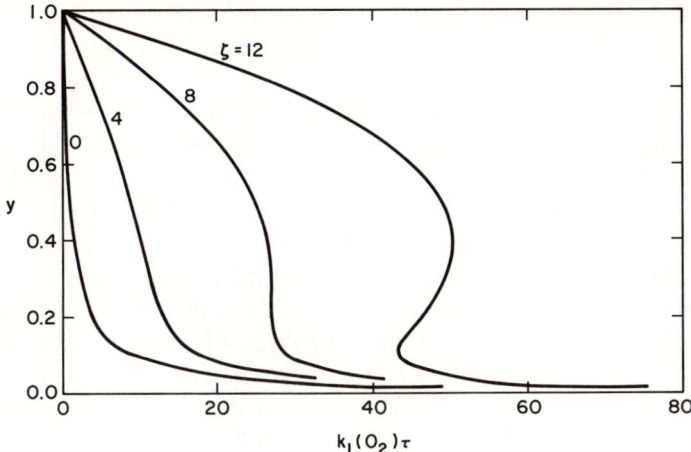

Figure 4. Conversion in CSTR

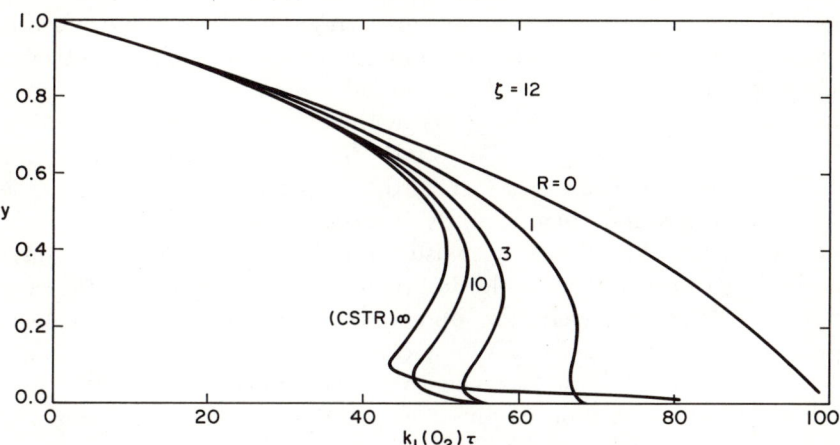

Figure 5. Conversion in recycle reactor

These theoretical advantages have been confirmed by the experimental data of Doelp et al. (51) and by Michalko (52). These observations can lead to a complete reversal of the conventional policy of optimal catalyst design: deposition of the catalyst on the very edge of the support to form the so-called "eggshell" catalyst should be replaced by deeper deposition to form "egg yolk" catalysts.

A highly exothermic reaction can give rise to multiple steady-state solutions, hysteresis, and instability in a reactor. Davies and Buben observed these instabilities in the oxidation of hydrogen and ammonia on platinum wires (53). Highly self-poisoned bimolecular Langmuir kinetics can give rise to all these instability phenomena even in an isothermal system.

Cardoso and Luss (43) have studied the oxidation of butane and CO over coiled platinum wires. When the temperature of the gas was raised and then lowered, the temperature of the wire demonstrated two branches of steady-state solutions with hysteresis. These instabilities are associated with reaction heat effects since the temperature difference between the wire and the gas can be several hundred degrees. Hugo and Jakubith (44) observed multiple steady-state conditions and hysteresis with CO oxidation on a wire mesh of platinum. They believed the system to be isothermal and hypothesized the cause of oscillations as an alternation of adsorbed CO between a linear and bridge form.

Luss and Amundson (54) have pointed out that for isothermal reactions inside a porous catalyst, multiple steady states cannot take place if the maximum rate occurs at the surface. Luss and Lee (55) computed multiple steady states in an isothermal catalyst slab when the diffusivities of the two reactants are markedly different and when the reaction rate depends on one reactant positively and on the other reactant negatively. A case of isothermal instability was observed by Beusch, Fieguth, and Wicke (56) with CO oxidation over platinum deposited on alumina pellets of 3-mm diameter. As the concentration of CO in the gas phase was increased, the production rate of CO_2 initially increases and then abruptly descends to a lower branch; when the CO concentration was decreased, the CO_2 production rate stayed on the lower branch for a while, and then abruptly ascended to the upper branch. Oscillations were also observed. Their data can be explained qualitatively by the theory of Wei and Becker (49).

The peculiarities of this negative order kinetics has led to many re-evaluations in reaction engineering practices, and they are important in designing the catalytic mufflers. The instabilities may be related to the sudden monolith melting and to the phenomenon of "breakthrough" —sudden transient emission of unconverted CO and hydrocarbons in a a fully warmed-up converter.

Transience and Dynamics

Most industrial catalytic reactors are designed to operate within a very narrow range of inlet temperatures, flow rates, and concentrations. Except for start-up, a change in feedstock, or an occasional upset, steady-state conditions prevail. On the other hand, rapid transience and wide dynamic ranges are the rule with the catalytic muffler. The proper engineering of the catalytic muffler requires an accurate transience analysis and an optimal reactor design over a wide range of input variables with aging catalysts.

The required level of conversion over the entire Federal Cycle is in the range of 70–90%. Since the catalytic converter is not hot enough during the first two minutes to be effective, the conversion after the first two minutes must be much better than the average value of 90%. A converter that effects better than 90% conversion at a high flow rate and low temperature with an aged catalyst must be overdesigned during a low flow rate and high temperature with fresh catalyst. Homogeneous gas phase oxidation may become important at temperatures above 1600°F.

Another area where transient analysis is sorely needed is destructive melting of monoliths, which occurs above 2500°F. These temperatures cannot be achieved by the sensible heat of the exhaust gases alone and must involve combustion of unburned CO and hydrocarbons in the exhaust gas. Thermal radiation must play a central role in the heat balance leading to such damage. In the absence of transport effects, the maximum temperature attained in a combustible mixture is the adiabatic flame temperature. However, if the combustion takes place exclusively on the wall, the maximum wall temperature is governed by the ratio of mass transfer coefficient of the combustible to the heat transfer coefficient. When the value of the Lewis number $c_p \rho D/\lambda$ is greater than 1, which is the case for hydrogen, the maximum wall temperature may exceed the adiabatic flame temperature. This phenomenon was predicted by Heck (57) and experimentally confirmed by Hegedus (58).

Vardi and Biller (59) are the first to realize the central importance of modeling the reactor bed temperature during a cold start. Their model of the reactor is one dimensional with separate temperatures for gas and solid but without longitudinal dispersion and without chemical reaction. To perform the computation, they replaced the continuum in the longitudinal direction with finite intervals which reduces the reactor to a series of 10 to 30 tirred tanks. Their input conditions derive from the seven mode California Cycle of the 1960's. They concluded that most of the transience takes place in the first two minutes and that the temperature between the gas and the solid is seldom higher than 50°F.

The first successful model that describes the complete performance of a catalytic muffler during a cold-start Federal Cycle was given by Wei (17) and by Kuo et al. (18). It describes a radial-flow pellet bed with base metal oxide eggshell catalyst, by using a Cascade model where each cell comprises three to four rows of pellets. A complete history of the inlet gas flow rates, temperatures, and gaseous compositions entering the catalytic muffler constitutes the input. The temperature and gaseous concentrations in each cell are computed as functions of time. This model has been used to predict muffler outlet concentrations and bed temperatures which fit the experimental data accurately. This model has been extensively used to explore the relative importance of numerous variables of design and operation and to point to promising avenues for new design.

Harned investigated the modeling of monolithic beds with base metal and precious metal catalysts (60). The bed is assumed to be heterogeneous piston flow without axial dispersion, and the kinetic used is simple negative first order. Finite difference methods were used so that the axial length is effectively divided into 10 to 40 zones. This creates a Cascade model, where the number of cells is assumed to be independent of a tenfold change in flow rate. No attempt was made to compare these calculations with experimental data, and the validity of this model is yet to be established. Ferguson and Finlayson have demonstrated that quasi-static solutions are close to dynamic solutions (61). Bauerle and Nobe also modeled pellet beds as heterogeneous piston flow without axial dispersion, again without experimental confirmation (62). Young and Finlayson challenged the validity of these monolithic models for neglecting radial gaseous temperature gradient (33). They showed that the Nusselt and Sherwood numbers cannot be considered passive parameters that are assigned ahead of time since they depend on the rate of chemical reaction. Our great need today, as always, is a rapid, closed loop between theoretical predictions and experimental testing.

The durability of the catalyst is limited by thermal aging and by poisoning (63). The deposition of lead and sulfur in a monolith is heavily concentrated at the reactor inlet (50). A similar tendency is usually not observed in a packed bed, possibly because of a slow catalyst circulation driven by the pulsating gas flow. Lead and sulfur penetrations inside a porous support are mainly confined to a depth of 0.6 mm. The optimal reactor design for such aging characteristics is a fertile field for imaginative optimization.

Conclusion

This review has described the catalytic muffler as:
new—lots of room for improvement,
big—annual sales of several hundred million dollars,
here to stay—no effective competition till 1980's.

A few past achievements are dwarfed by the many unsolved problems which pose a challenge we cannot ignore.

Nomenclature

a	surface to volume ratio
C	pollutant concentration
c_p	heat capacity
D	dispersion coefficient
d	hydraulic diameter of channel
d_p	diameter of pellets
F	flow rate of gas
H	heat of reaction
h	heat transfer coefficient between gas and solid
k	mass transfer coefficient between gas and solid
L	length of bed
m	molecular weight of pollutant
N	number of cells in Cascade model
n	reaction order
Nu	Nusselt number, hd/λ
Pe	Peclet number ud/D and $ud\rho c_p/\lambda$
Pr	Prandtl number $c_p\mu/\lambda$
Q	activation energy
R	reaction speed
R_g	gas constant
r	distance in radial direction
r_o	radius of tube
Re	Reynolds number $du\rho/\lambda$
Sc	Schmidt number, $\mu/\rho D$
Sh	Sherwood number, kd/D
T	temperature
TU	transfer units, Nu$\cdot aL/$Re$\cdot$Pr or Sh$\cdot aL/$Re$\cdot$Sc
t	time
u	gas velocity in bed
V	vehicle speed
W	gm/mile of pollutant emission in Federal Cycle
x	distance, longitudinal direction
α	thermal diffusivity, $\lambda/\rho c_p$ in gas
β	eigenvalue in Graetz problem
λ	heat conduction coefficient
μ	viscosity
ζ	$k_2 C$ inlet
ρ	density
τ	residence time
ψ	eigenfunction in Graetz problem

Literature Cited

1. "Report of the 65th General Motors Stockholders Meeting," Detroit, Mich., May 25, 1973.
2. "Evaluation of Catalysts as Automotive Exhaust Treatment Devices," Panel Report of the Committee on Motor Vehicle Emission, National Academy of Sciences, March 1973; "Evaluation of Catalytic Converters for Control of Automobile Exhaust Pollutants," Sept. 1974.

3. Wei, James, "Catalysis for Motor Vehicle Emissions," *Advan. Catalysis* (1975) **24**, 57.
4. *Federal Register* **35** (219) Nov. 10, 1970; **36** (128) July 2, 1971.
5. "Report by the Committee on Motor Vehicle Emissions," National Academy of Sciences, Washington, D.C., Feb. 15, 1973.
6. Williamson, S. J., "Fundamentals of Air Pollution," Chap. 8, Addison-Wesley, 1973.
7. Pierson, W. R., Hammerle, R. H., Kummer, J. T., *Soc. Auto. Eng.* (1974) paper 740287.
8. "An Evaluation of Alternative Power Sources for Low Emission Automobiles," Panel Report of the Committee on Motor Vehicle Emissions, National Academy of Sciences, April 1973.
9. *Oil Gas J.* (June 10, 1974) p. 105.
10. Burke, D. P., *Chem. Week* (Nov. 1, 1972) p. 23.
11. Shinnar, R., *Science* (1972) **175**, 1357.
12. "Feasibility of Meeting the 1975-76 Exhaust Emission Standards in Actual Use," Panel Report of the Committee on Motor Vehicle Emission, National Academy of Sciences, June 1973.
13. Kramers, H., Westerterp, K. R., "Elements of Chemical Reactor Design and Operation," Chap. 3, Netherlands University Press, 1963.
14. Yagi, S., Kunii, D., Wakao, N., *AIChE J.* (1960) **6** (4) 543.
15. Votruba, J., Hlavacek, V., Marek, M., *Chem. Eng. Sci.* (1972) **27**, 1845.
16. Froment, G. F., ADVAN. CHEM. SER. (1972) **109**, 1.
17. Wei, J., *Chem. Eng. Progr., Monograph Ser.* (1969), Ser. **6**, No. 65.
18. Kuo, J. C. W., Morgan, C. R., Lassen, H. G., *Soc. Auto. Eng.* (1971) paper 710289.
19. Votruba, J., Hlavacek, V., *Chem. Eng. J.* (1972) **4**, 9.
20. Hlavacek, V., Kubicek, M., *Chem. Eng. Sci.* (1972) **27**.
21. Lemme, C. D., Givens, W. R., *Soc. Auto. Eng.* (1974) paper 740243.
22. Howitt, J. S., Sekella, T. C., *Soc. Auto. Eng.* (1974) paper 740244.
23. Eckert, E. R. G., Drake, R. M., Jr., "Analysis of Heat and Mass Transfer," Chap. 7, McGraw-Hill, New York, 1972.
24. Kays, W. M., London, A. L., "Compact Heat Exchangers," 2nd ed., McGraw-Hill, New York, 1964.
25. Shah, R. K., London, A. L., Technical Report No. 75, Contract 225 (91) for Office of Naval Research, Dept. of Mech. Eng., Stanford University, 1971.
26. Sherony, D. F., Solbrig, C. W., *Int. J. Heat Mass Transfer* (1970) **13**, 145.
27. Schoenherr, R., Woolley, J. A., Chow, W., presented ACS, April 1974.
28. Heck, R. H., Wei, J., Katzer, J. R., ADVAN. CHEM. SER. (1974) **133**, 34.
29. Koch, H., Ph.D. Thesis, Technical University of Munich, 1973.
30. Katz, S., *Che.m Eng. Sci.* (1959) **10**, 202.
31. Solomon, R. L., Hudson, J. L., *AIChE J.* (1967) **13**, 545.
32. Lupa, A. J., Dranoff, J. S., *Chem. Eng. Sci.* (1966) **21**, 861.
33. Young, L. C., Finlayson, B. A., ADVAN. CHEM. SER. (1974) **133**, 629.
34. Vortmeyer, D., Schaefer, R. J., *Chem. Eng. Sci.* (1974) **29**, 485.
35. Hegedus, L. Louis, "Abstracts of Papers," 166th National Meeting, ACS, Chicago, 1973, PETROL 008.
36. Johnson, W. C., Chang, J. C., *Soc. Auto. Eng.* (1974) paper 740196.
37. Morikawa, K., Benedict, W. S., Taylor, H. S., *J. Am. Chem. Soc.* (1936) **58**, 1795.
38. Sinfelt, J. H., *J. Catalysis* (1972) **27**, 468.
39. Langmuir, I., *Trans. Faraday Soc.* (1922) **17**, 621.
40. Schwab, G. M., Gossman, K., *Z. Phys. Chem.* (1958) **16**, 39.
41. Tajbl, D. G., Simons, J. B., Carberry, J. J., *Ind. Eng. Chem., Fundamentals* (1966) **5**, 171.

42. Sklyarov, A. V., Tretyakov, I. I., Shub, B. R., Roginskii, S. Z., *Dokl. Akad. Nauk. SSSR* (1969) **189**, 1302.
43. Cardozo, M. A. A., Luss, D., *Chem. Eng. Sci.* (1969) **24**, 1699.
44. Hugo, P., Jakubith, M., *Chem-Ing.-Tech* (1972) **44**, 383.
45. Cochran, H. D., Ph.D. Thesis, Chemical Engineering, MIT, 1972.
46. Voltz, S. E., Morgan, C. R., Liederman, D., Jacob, S. M., *Ind. Eng. Chem., Prod. Res. Develop.* (1973) **12**, 294.
47. Roberts, G. W., Satterfield, C. N., *Ind. Eng. Chem., Fundamentals* (1966) **5**, 317.
48. Satterfield, C. N., "Mass Transfer in Heterogeneous Catalysis," MIT Press, Cambridge, Mass., 1970.
49. Wei, J., Becker, E. R., ADVAN. CHEM. SER. (1975) **143**, 116.
50. MacArthur, D. P., ADVAN. CHEM. SER. (1975) **143**, 85.
51. Doelp, L. C., Koester, D. W., Mitchell, M. M. Jr., ADVAN. CHEM. SER. (1975) **143**, 133.
52. Michalko, E., U.S. Patent **3,259,589** (July 1966).
53. Frank-Kamenetskii, D. A., "Diffusion and Heat Exchange in Chemical Kinetics," translated by N. Thon, Chap. 9, Princeton University Press, 1955.
54. Luss, D., Amundson, N. R., *AIChE J.* (1967) **13** (2), 279.
55. Luss, D., Lee, J. C. M., *Chem. Eng. Sci.* (1971) **26** (9), 1433.
56. Beusch, H., Fieguth, P., Wicke, E., ADVAN. CHEM. SER. (1972) **109**, 615.
57. Heck, R. H., Ph.D. Thesis, Chemical Engineering, University of Delaware, 1974.
58. Hegedus, L. L., "Temperature Excursions in Catalytic Monoliths," AIChE Meeting, Washington, D.C., December 1974.
59. Vardi, J., Biller, W. F., *Ind. Eng. Chem., Process Design Develop.* (1968) **7**, 83.
60. Harned, J. L., *Soc. Auto. Eng.* (1972) paper 720520.
61. Ferguson, N. B., Finlayson, B. A., *AIChE J.* (1974) **20**, 539.
62. Bauerle, G. L., Nobe, K., *Ind. Eng. Chem., Process Design Develop.* (1973) **12** (4), 407.
63. Hegedus, L. L., *Ind. Eng. Chem., Fundamentals* (1974) **13**, 190.

RECEIVED December 4, 1974. This review was written with the support of the National Science Foundation grant GK 38189 and the support of the Minnesota Mining and Manufacturing Co.

2

Kinetic Models in Heterogeneous Catalysis

SOL W. WELLER

Department of Chemical Engineering, State University of New York at Buffalo, Buffalo, N. Y. 14214

The diverse meanings of "kinetic models" and the varied objectives in establishing models are reviewed. Three commonly used and theoretically derivable isotherms—Langmuir, Freundlich, and Temkin—attempt to relate surface concentrations to gas phase composition. In the Langmuir isotherm, problems arise with enhanced adsorption from reactive mixtures, apparent negative values of adsorption coefficients, and abnormal temperature dependence of adsorption coefficients. Various approaches in the past 10 years have been taken to kinetic modelling in heterogeneous catalysis. A few procedural suggestions are offered for those whose primary interest is reactor design and for those whose primary interest is reaction mechanism.

The subject of kinetic models in heterogeneous catalysis is a troubling one. A rigorous theory of heterogeneous catalytic kinetics is analogous to a Greek tragedy: the protagonist is doomed to failure, and the course of that failure is known to the observers from the start. The roots of this preordained doom for practical systems are two-fold.

First, the surface of a practical catalyst is not uniform, geometrically or energetically; the adsorbate molecules may interact both to produce surface complexes and also to give adsorption energies that depend on surface coverage; neither of these effects is quantitatively predictable today from first principles for any arbitrary system. As a result, we have no way of knowing in advance, for any system, which of several different isotherms is quantitatively valid. More is said about different isotherms later, and I point out here only that each of them depends on the unverifiable assumption of some model about the uniformity of the surface or the distribution of energies.

The second root of the tragedy lies in the need to use adsorption isotherms at all. In homogeneous kinetics, and often in enzymatic kinetics, one can measure directly the concentrations of the reacting species. With the added assumption that the mass action law is applicable to elementary reaction steps, we can then piece out a reaction mechanism which can be examined for consistency with the global kinetics. In heterogeneous catalysis we generally are ignorant of the surface concentrations. Even as powerful a tool as measurement of the IR absorption spectra of adsorbed species, for example, may give information about entities which are present in large concentration on the surface but may be irrelevant to the reaction mechanism. [The concept of a "most abundant surface intermediate" (masi), discussed by Boudart in his recent and excellent review (1), is concerned with surface species that are, in fact, intermediates.] We then add to this problem of indirect estimate of surface concentrations the implicit assumption that the simple mass action law applies to elementary surface reactions. The literature is replete with claims of rate-limiting surface reactions that are third or fourth order—a result which would make a homogeneous kineticist tremble.

Nevertheless, the reactions occur, we can make some sense out of the kinetics, and we have some feeling about the general principles that undrly heterogeneous kinetics. What is the rational man to do? There has been a spectrum of possible approaches, all of which have advocates, and examples of which I will give. Near one end of the spectrum are those who correctly claim that fundamental understanding is clearly preferable to raw empiricism and who also believe that there is adequate reason to accept the quantitative validity of some isotherm (say, Langmuir-Hinshelwood) and of the mass action principle, and that it is possible to establish the reaction mechanism (or at least the rate-limiting step) through consistency with the observed kinetics. Near the other end of the spectrum are those who claim that none of the theories has quantitative validity and that the important thing in practice is to obtain the most convenient empirical kinetic equation that provides adequate fit to observations and that permits design of commercial reactors intended to operate within the range of parameters studies. The intermediate positions are numerous. One popular one, at least among chemists turned kineticists, is to assert that some isotherm, say Langmuir-Hinshelwood, gives correct qualitative insights although it may lack quantitative validity, that useful forms of the global rate expression may be deduced on the basis of this assumption, but that by no means is the reaction mechanism to be believed as proved simply from consistency with kinetic data. Another intermediate position, intrinsically attractive, is to say that we will progressively complicate (as needed) the simplest theoretical expression (*e.g.*, of the Langmuir-Hinshelwood type), incorporate those changes required by heterogeneity or interaction, and suffer as a necessary evil the resulting complications in the final rate law. The difficulty with this, of course, is that we don't know *a priori* what complications are quantitatively justified for our system.

Perhaps the best way to proceed here is to try to answer some fundamental questions.

(1) Some definition of terms is appropriate. Specifically, what do we mean by "kinetic model"?

(2) Why do we wish to establish the kinetic model? What are our objectives as engineers or engineering scientists?

(3) Since we normally cannot directly determine the surface concentrations of reacting species, what is the range of validity of the various adsorption isotherms that relate surface concentrations to the observable gas-phase partial pressures?

(4) To what extent can kinetics furnish unambiguous answers to certain fundamental questions concerning the nature of the catalyst surface and the detailed course of the surface reactions which are actually occurring?

Kinetic Models

There are at least two interpretations of this expression to be found in the literature:

(a) The actual mechanism, at the molecular level, by which the chemical reaction occurs.

(b) A convenient and reasonable representation of the reaction which, although not in general unique, is at least consistent with known data and permits both interpolation and some extrapolation.

[Knözinger *et al.* (2) have recently drawn even more explicit distinctions in the following definitions:

"Mechanistic model: a reaction scheme which can be interpreted as a possible molecular mechanism, the intermediate species and active sites of which can be observed directly or must be postulated on the grounds of experimental evidence.

Kinetic equation: a rate equation that is deduced for a given mechanistic model.

Kinetic model: a purely formal reaction scheme whose intermediates and active sites are not interpreted as any real chemical species.

Formal kinetic equation: a rate equation that is deduced for a given kinetic model."]

The difference between the two interpretations is not trivial since it is always desirable in science to distinguish between reality and plausibility or, in other words, between necessity and consistency. In one case, establishing the kinetic model is considered the equivalent of establishing a unique truth concerning the way in which the reaction actually occurs. In the other, only consistency and convenience are involved. Boudart (3) has expressed an extreme position on the entire matter: "'Mechanism' or 'model' can mean an assumed reaction network, or a plausible sequence of steps for a given reaction, or a postulated stereochemical path during the course of an isolated step. Since methods of investigation and goals are so utterly different in the study of networks, sequences and steps, the words 'mechanism' or 'model' should be avoided. They have acquired the bad connotation associated with irresponsible or vain speculation, largely to describe achievements that vary widely in sophistication."

At this point it is useful to ask why we wish to establish the kinetic model for a surface reaction. If the reason is to prove conclusively the molecular mechanism by which the reaction is truly occurring on the catalyst surface, then the response must probably be that (1) the mechanism of no solid-catalyzed reaction is adequately understood at the molecular level, even for as simple a case as H_2–D_2 exchange, and (2) of all the sophisticated approaches that have been applied to the establishing of catalytic reaction mechanisms, kinetics is among the least useful in providing unambiguous answers. Much more useful, in different systems, are techniques such as isotopic tracer studies, isotopic rate effects, determination of stoichiometric number, investigation of the stereochemistry of complex reaction products, and infrared absorption spectra of adsorbed species, which have all given some insight as to the reaction mechanism.

On the other hand, if we wish a kinetic model for less ambitious reasons—to be consistent with rate data, to permit reactor design, to suggest new experiments based on predictions of the model, to contribute qualitative insight into a possible reaction path—then many approaches are possible and plausible.

Adsorption Isotherms

Most of us go through the following steps implicitly or explicitly, in deriving a rate equation on the basis of a particular kinetic model for a surface-catalyzed reaction:

(a) Choose a particular surface reaction as the rate-limiting step. (For simplicity, this discussion is limited to cases where such a procedure is justifiable—*i.e.*, where all heat and mass transfer steps and all adsorption and desorption processes are relatively rapid; only one step in an overall sequence is rate limiting, etc.) As an example, in a reaction $2A \rightarrow B + C$, we may wish to consider the implication of assuming that (1) reaction occurs only between molecules of chemisorbed A, and (2) the rate-limiting step is reaction between two molecules of A adsorbed on adjacent sites.

(b) Assume that a conventional mass-action law applies to such surface reactions, where the Guldberg-Waage "active masses" are proportional to the surface concentrations (or fractional coverage).

(c) Assume that some isotherm equation correctly relates the surface concentration of any species to the observable partial pressures of all species in the ambient bulk gas.

(d) Deduce the corresponding rate equation relating the global kinetics to the observable partial pressures.

Although the following discussion will focus largely on step c, brief comments about a and b are appropriate. First, it is common to treat surface reactions in terms of the fundamental concept advanced by Langmuir (*4, 5*) and Hinshelwood (*6*)—i.e., that the molecules react while adsorbed, with coverage not exceeding a monolayer. Langmuir himself, in his extraordinary 1921 paper on the Pt-catalyzed oxidation of CO and H_2 (*5*), suggested an alternate possibility: that reaction may occur directly on collision of a gas molecule with an adsorbed molecule or atom. This concept was later developed by Rideal and Eley (*7, 8*) and has been variously labelled the Langmuir-Rideal, Rideal-Eley, Eley-Rideal, or "dive-bomb" mechanism. The resultant rate equations differ significantly from the Langmuir-Hinshelwood form. There are few, if any, unambiguous examples where this mechanism is operative. [Nevertheless, Sinfelt has had great practical success in applying a mildly complicated version of this mechanism in the treatment of hydrocarbon hydrogenolysis over supported metals. His model is derived from that used earlier by Taylor and co-workers at Princeton. *cf.* Ref. *10.*] However, the Mars-van Krevelen (*9*) approach to oxidation reactions over transition metal oxides does in fact invoke the notion of cyclic redox reaction of the catalyst directly on collision of a gaseous molecule (substrate or O_2) with the (oxidized or reduced) surface. This approach has achieved considerable popularity; its fundamental formulation and limitations are considered in some detail later.

With respect to item b, if the binding energies of adsorbed molecules are not identical, there is no reason to expect equal reactivity of those molecules. The direct result is that the assumption of a surface mass-action rate law is without theoretical justification for systems of non-uniform energetics. To put the matter another way, the surface rate constant k, normally assumed to be only a function of temperature, will also be a function of surface coverage in the general case.

We now consider item c and inquire: what is the proper equilibrium adsorption isotherm to be used in developing a rate equation? Again for simplicity, the discussion is limited to adsorption without dissociation. Before attempting an answer, we review some salient characteristics of the Langmuir, Freundlich, and Temkin isotherms and the assumptions involved in their derivation. These are singled out because of their relevance to monolayer chemisorption.

Langmuir Isotherm. The Langmuir isotherm, initially derived on the basis of kinetic arguments (*11*), can be derived equally well from the statistical thermodynamics appropriate to equilibrium in ideal localized monolayers (*12*). However, in either case the validity of the simple isotherm depends on the satisfaction of a set of postulates, some of which are more easily granted than others as being true for real systems. Regardless of any questions concerning the quantitative validity of all the postulates, Langmuir's concepts are of fundamental importance to all subsequent theoretical work in the field.

Some of the key assumptions and results of the elementary equation are:

(1) Adsorption occurs on a finite number of equivalent sites on a uniform surface.

(2) Each site can adsorb one and only one gas molecule, and the adsorption is localized. This implies localized monolayers. If the molecules were totally free to move over the surface with no mutual interaction, they would behave as a two-dimensional perfect gas. Freedom to move, but with varying degrees of mutual interaction, would correspond to the appropriately imperfect, two-dimensional gas.

(3) The adsorbed molecules do not interact, and their energies are independent of the presence or absence of adsorbed molecules on neighboring sites.

(4) If two or more species of gas molecules are present, they will compete for adsorption on the fixed number of equivalent sites.

(5) For any gaseous species i, the corresponding equilibrium adsorption constant K_i is the ratio of an adsorption rate constant k_{i_a} and a desorption rate constant k_{i_d}. K_i is intrinsically positive. Moreover, $K_i = \exp[-\Delta G_i^\circ/RT] = \exp[+\Delta S_i^\circ/R] \exp[-\Delta H^\circ_i/RT]$, where ΔG_i, ΔS_i°, and ΔH_i° are the free energy, entropy, and enthalpy change on adsorption.

(6) For any gaseous species i at a partial pressure p_i, the fraction of sites covered by adsorbed i (θ_i) increases linearly with p_i at sufficiently low values of p_i; conversely, θ_i becomes independent of p_i at sufficiently high values of p_i.

The problems with the quantitative, literal application of the simple Langmuir isotherm have been presented in many places. Only a few additional comments are made here, chiefly in connection with the temperature dependence and the sign of the constants K_i. Parenthetically, Langmuir was one of his own most severe critics. In his 1938 Faraday Lecture (13) he observes, in discussing his work on the adsorption of cesium on tungsten, ". . . the physical assumptions underlying this factor $(1 - \theta)$ [in the equation for the adsorption rate] are very improbable. . . . Even if there were no mobility at all, we cannot justify the factor . . . if the sites are closely adjacent to one another. In this case it is more reasonable to assume that $(1 - \theta)$ should be replaced by $(1 - \theta^4)$."

Boudart, in 1956 (14) and since, has properly emphasized the importance of examining the temperature dependence of the equilibrium adsorption constants K_i in order to establish that the heats and entropies of adsorption are reasonable. He quoted the example of stibine decomposition on antimony, where the kinetic data are well fitted by the Langmuir equation for a single site model, $r = kKp/(1 + Kp)$. Experimentally, K is essentially independent of temperature, implying that $\Delta H \simeq 0$, an unreasonable result. The deduced ΔS_a is also unreasonable. Boudart's suggestion that the "paradox of heterogeneous kinetics" can account for the temperature independence of K involves the idea that, although the surface coverage decreases with increasing temperature, sites of higher adsorption energy (in an energetically heterogeneous system) become operative at higher temperature. The qualitative argument is plausible, but there are difficulties in developing quantitative rigor: (a) the actual distribution of energetics is always quantitatively unknown; and (b) it would be coincidental that these two effects of opposite sign should happen to just balance.

Another example illustrating strange behavior of an adsorption constant is the fascinating disproportionation of propylene to butene and ethylene over tungsten oxide–silica. Both Luckner et al. (15) and Hattikudur and Thodos (16) agree that a dual-site surface reaction is rate controlling and that the initial rates are given by the standard dual-site Langmuir-Hinshelwood equation:

$$r_o = C p_3^2 / (1 + K_3 p_3)^2$$

where K_3 and p_3 are the adsorption constant and partial pressure, respectively, of propylene. The work of Hattikudur and Thodos is particularly convincing because of the unusually large range of partial pressures investigated (more than 30-fold for propylene).

The unexpected result is that, as both sets of investigators agree, K_3 actually increases with increasing temperature whereas one normally expects a decrease. Hattikudur and Thodos report a value of $\Delta H_a = +12,140$ cal/g-mole for the adsorption of propylene. (The data of Luckner et al. do not fall on a straight line for $\ln K_3$ vs. $1/T$.) Now endothermic adsorption is possible (17), but there are very few authenticated examples and these seem to involve either molecular hydrogen or molecular oxygen (18). Since the free energy of adsorption must be negative, endothermic adsorption implies that $T\Delta S_a > \Delta H_a$. Thomas and Thomas suggest, for heuristic purposes, two ways of having an entropy increase on adsorption rather than the expected decrease. The first invokes dissociative adsorption with complete two-dimensional mobility of the adsorbed fragments; the second is that for some reason the entropy of the solid itself increases more than the entropy of the adsorbed gas decreases. For present purposes we note only that the first suggestion would mean giving up the Langmuir isotherm, for which immobile adsorption is postulated; moreover, the picture seems intuitively improbable for propylene disproportionation. The second suggestion raises a difficult theoretical problem, which I believe no one has quantitatively attempted to date.

The last point to be mentioned relates to the sign of the K_i's. In the standard application of Langmuir-type isotherms, those kinetic models are discarded that lead to rate equations for which any K_i determined from experimental rate data turns out to be significantly negative. Some years ago it was noted (19) that the limited work available on mixed adsorption from a potentially reactive gas mixture indicated that enhanced, rather than competitive, adsorption might be a common phenomenon and that force-fitting of "enhanced adsorption" data into a Langmuir form guarantees a negative value for at least one K_i in the range of conditions studied. In the intervening years more examples of adsorption from reactive mixtures have been examined, and the enhanced adsorption behavior for at least one (sometimes both) of the constituents does seem to be the general pattern. The behavior of H_2–CO mixtures on Co and Fe Fischer-Tropsch catalysts, ZnO, and noble metals has been recently reviewed by Gupta *et al.* (20). Gerai *et al.* (21), in a study of adsorption from C_3H_6–O_2 on CuO and Cu_2O, report enhanced adsorption of O_2 from the mixture and decreased adsorption of C_3H_6; at all temperatures and from all mixtures the total adsorption was more than additive for both CuO and Cu_2O.

Intuitively we can understand the reason for enhanced adsorption, when it occurs, in terms of the formation of a surface complex which is more strongly adsorbed than either component singly. Direct evidence, from IR spectra, calorimtery, and other techniques, is available for such complexes in a number of cases (*see*, for example, Refs. 21–24).

It has been suggested that if all the other criteria required for quantitative validity of the Langmuir-Hinshelwood theory were satisfied, we could take into account explicitly the formation of such surface complexes, with the addition of another equilibrium constant, and maintain the form of the Langmuir expressions. Unfortunately, it is difficult experimentally to establish the existence, nature, and surface equilibrium constants for such complexes in a quantitative way, and very few kineticists interested in mechanism have taken the trouble to investigate this problem.

Freundlich and Temkin Isotherms. Since the several derivations of the Freundlich and Temkin isotherms are summarized in various reports (25, 26, 27, 28), the details are not repeated here. Readers not familiar with Sips' elegant paper may be interested to know that he has used Stieltjes transforms to solve the inverse of the usual problem—*i.e.*, to deduce the possible distribution functions for adsorption energy that are consistent with the experimental isotherm $\theta(p)$.

All derivations of these isotherms utilize the Langmuir result for adsorption on surface sites of one particular adsorption energy, and in this sense the Langmuir isotherm is a fundamental starting place. I do wish to make a few points:

(1) The Freundlich and Temkin isotherms are no less (and no more) theoretically justified than the Langmuir isotherm. The basic concept of localized, monolayer chemisorption is required for all. Different isotherms result according to the assumptions made about the heat of adsorption, ΔH_a, and the site distribution function, $N(\Delta H_a)$, among other things. The Langmuir isotherm requires the assumption that $-\Delta H_a$ = constant. The Freundlich isotherm can be deduced from the assumptions that $N = ae^{-\Delta H/\Delta H_0}$ and that θ (or p) is small; [cf. the comment by Thomas and Thomas (Ref. 18, p. 44): "The Freundlich equation is . . . no longer to be regarded as merely a convenient form of representing the Langmuir equation at intermediate values of θ. Moreover, the method of derivation disposes of the criticism that the Freundlich equation predicts a progressively increasing coverage with increasing pressure: the isotherm is expected to be valid only at low coverages."] The Temkin equation is obtained if the adsorption heat decreases linearly (between maximum and minimum values) with surface coverage θ and if θ is in a middle range.

(2) We do not know, for any arbitrary system, which assumptions about the distribution functions for energy are true or which isotherms are therefore theoretically valid. [In this context, Clark (Ref. 26, p. 57) has clearly stated a dual difficulty: "It should be emphasized . . . that agreement between the theoretical isotherms with distribution functions determined by the various procedures . . . and experimental isotherms does not guarantee that the true physical picture has been discovered. . . . Another difficulty is the inherent insensitivity of the theoretical isotherm to the form of the distribution function within the accuracy of experimental data."]

(3) Restraint is therefore appropriate in insistence that any one isotherm has unique quantitative validity in applications to catalytic kinetics.

Rate Equations and Kinetic Models: An Eclectic Review

This section contains a sampling from the enormous literature on catalytic kinetics published during the last 10–15 years. The samples were chosen partly to illustrate the spectrum of approaches used to interpret kinetic data and partly to indicate some areas of substantial disagreement.

Ammonia Synthesis: an Example of Ambiguity. Boudart's 1972 review (1) contains an excellent discussion of this system as exemplifying "two-step catalytic reactions." My purpose here is only to add to the variously proposed rate equations some recent results of Brill (29) and to indicate the difficulty, even in this most-studied of catalytic reactions, of drawing conclusions about such basic questions as (a) is the surface of an iron catalyst uniform or heterogeneous, and (b) if the surface is uniform, is H_2 adsorbed molecularly on a single site or dissociatively on dual sites.

The classic Temkin-Pyzhev equation uses the Temkin isotherm (implying a heterogeneous surface) for the surface concentration of nitrogen as related to the fictitious nitrogen partial pressure that would be in equilibrium with the actual hydrogen and ammonia partial pressures. For the rate of the forward reaction, the Temkin-Pyzhev treatment gives:

$$r = k\, p_{N_2} \left(\frac{p_{H_2}^{3}}{p_{NH_3}^{2}}\right)^i$$

The constant i must be determined; to agree with experiment, it is often chosen to be between 0.5 and 0.6.

The 1942 paper of Brunauer et al. (27), which is more explicit in its derivations, also starts with the assumption of linear decrease in adsorption heat with increasing coverage (heterogeneity) and ends with a generalized "Temkin isotherm" that is valid for the entire range of adsorption, whereas the original Temkin isotherm is valid only in the middle range. Dissociative adsorption of N_2 is again assumed, as in Temkin-Pyzhev. For the forward reaction only the rate expression at low surface coverage (θ_N) becomes:

$$r = \frac{k\, p_{N_2}}{\left(1 + \dfrac{a_o\, K\, p_{NH_3}}{p_{H_2}^{1.5}}\right)^{2\alpha}}$$

If $\alpha = 1$ (arbitrary choice), this reduces to:

$$r = \frac{k\, p_{N_2}}{\left(1 + \dfrac{a_o\, K\, p_{NH_3}}{p_{H_2}^{1.5}}\right)^{2}}$$

Boudart has shown (14, 30) that the assumption of a homogeneous surface (Langmuir model) and dissociative adsorption of nitrogen as the rate-determining step leads to the synthesis equation:

$$r = \frac{k\, p_{N_2}}{\left(1 + \dfrac{b\, p_{NH_3}}{p_{H_2}^{1.5}}\right)^{2}}$$

which is identical with the equation deduced by the procedure of Brunauer et al. for a heterogeneous surface (provided $\alpha = 1$).

Still more recently, Brill and co-workers have reported interesting results:

(a) Field electron microscopic studies show that N_2 adsorption occurs preferentially on (111) faces of iron and very much less on (100) and (110) faces (31).

(b) On the basis of field ion mass spectrometric studies, N_2H appears to be the first product formed on exposure of iron to N_2–H_2 (*32*).

(c) IR studies of Fe–MgO catalyst exposed to N_2–H_2 at 410°C indicate the presence of partially hydrogenated N_2 molecules (*i.e.*, undissociated) having bonds similar to those hydrazine (*33*).

On this basis, Brill derived a rate expression for ammonia synthesis under the assumptions that the surface is homogeneous but that the rate-determining step is the non-dissociative adsorption of molecular nitrogen on a single site (*29*). Furthermore, he shows that a set of experimental integral conversion data taken at 340°C is satisfied by this rate expression with a standard deviation of ± 1.5% in the rate constant. The same data, fitted by the equation of Ozaki *et al.* for dissociative adsorption on dual sites, gives an almost identical standard deviation, ± 1.4%, in the corresponding rate constant.

Hopefully further ingenious experiments will shed further light on the reaction mechanism (although after 40 years one may approach pessimism) but in this special case, at least, the fitting of global rate expressions seems unlikely to differentiate kinetic models that differ radically in their assumptions.

Para–Ortho-Hydrogen Conversion: an Example of Simplicity. The paramagnetic mechanism for the para–ortho-hydrogen shift reaction at low temperatures should constitute the simplest chemical example of a catalyzed reaction. Hutchinson *et al.* (*34*) conducted a careful study of the approach to equilibrium, from both sides of the equilibrium composition, over ferric oxide gel at 76°K. The results are disconcerting. They may be summarized as follows:

(1) With the Langmuir-Hinshelwood approach, the overall rate expression has the same form regardless of which step (adsorption, surface reaction, or desorption) or combination of steps is rate limiting. This simplifies critical testing of the approach.

(2) The deduced Langmuir-Hinshelwood rate constant is different depending on whether one approaches equilibrium from the ortho-rich or para-rich side. This result would violate the principle of microscopic reversibility and indicates that no simple Langmuir-Hinshelwood model can be applicable in this system.

(3) The discrepancy between theory and experiment can be resolved qualitatively if one postulates that the activation energies for adsorption and desorption change with coverage of the catalyst surface. However, such a postulate is tantamount to giving up the quantitative validity of the Langmuir-Hinshelwood model.

Dehydration of Ethanol to Ether. One of the most gratifying publications in catalytic kinetics is the paper of Kabel and Johanson (*35*) on the vapor-phase dehydration of ethanol to diethyl ether over the acid form of Dowex 50—a sulfonated styrene–divinylbenzene copolymer. The authors found their kinetic data to be consistent with the equation predicted from a simple Langmuir model in which the rate-limiting step is a surface reaction between adjacently adsorbed ethanol molecules:

$$r = \frac{k_s {}^*K_A{}^2 \left[P_A{}^2 - \dfrac{P_E P_W}{K_{eq}}\right]}{[1+K_A P_A + K_W P_W + K_E P_E]^2}$$

the subscripts A, W, and E representing ethanol, water, and ether, respectively. This itself is plausible, but it is not the reason for the exceptional interest of the work. The authors proceeded to do that which is seldom done—namely, to determine K_A, K_W, and K_E directly and independently from adsorption isotherms (Langmuir) of the individual pure components. The extraordinary result was the agreement between the two sets of adsorption constants—one set determined by fitting of the kinetic data, the other set from the individual isotherms. At 120°C, for example, K_A, K_W, and K_E determined from the kinetics were 3.4, 7.0, and ~ 0. The corresponding values from the isotherms were 2.5, 7.6, and ~ 0.

The experience with Dowex 50 does not permit extension to another common catalyst for alcohol dehydration—alumina. Knözinger and Pines have conducted the most extensive research on the mechanism of alcohol reactions over alumina. Knözinger et al. (2) have recently analyzed earlier data (36) on the kinetics of ether formation by a grid search method. The experimental data could not be fitted by equations based on the expected mechanism. On the other hand, five formal kinetic equations were developed which describe the data equally well (with mean error identical to the mean experimental error) and which cannot be distinguished. All five give comparable activation energies. The authors conclude: "At least for the ether formation on alumina. . . . , it therefore seems unrealistic to use kinetic analysis for the elucidation of the molecular mechanism of the reaction. The only possible result is a purely formal description of the reaction rate as a function of the partial pressures of alcohol, ether, and water."

Hydrogenation of Cyclopropane. A second example of a reaction in which agreement is claimed between the adsorption constant deduced from a rate equation (Langmuir form) and that measured from the adsorption isotherm is the recent paper of Sridhar and Ruthven (37). The kinetics for cyclopropane hydrogenation over four supported nickel catalysts at 60°C were found to be best fitted by the simple expression:

$$r = k\, K_c p_c\, (1+K_c p_c)^{-1}$$

where subscript c represents cyclopropane. The values of the kinetically determined K_c for the four catalysts ranged from 2.12 to 3.71 atm^{-1}; the value of K_c calculated from the isotherm data of Benson and Kwan (*38*) was reported to be 2.5 atm^{-1}, in good agreement.

Several aspects of the work lead one to be less than totally sanguine about the results:

(1) The actual plots of r vs. p_c are sigmoidal—a result which, as the authors appreciated, is not truly compatible with the above rate equation.

(2) No term involving p_H appears in the "adsorption denominator" despite the fact (which the authors note) that hydrogen is more strongly adsorbed on nickel than is cyclopropane.

(3) The commercial Ni–SiO$_2$–Al$_2$O$_3$ catalyst studied by Benson and Kwan was obtained from a different source (Atlantic Refining Co., *vs.* Harshaw).

(4) Benson and Kwan fitted their own rate data for cyclopropane hydrogen to a power rate law, fitted their own adsorption data to Freundlich isotherms and proposed that the rate-determining step was a surface reaction between adsorbed cyclopropane and an adsorbed hydrogen atom.

Statistical Model Building. In 1962 a seminal paper by Box and Hunter (*39*) described an extremely powerful, iterative model-building method involving minimization of the residuals of a diagnostic parameter. Hunter and Mezaki (*40*), in applying this approach, discussed the results of an experimental study of methane oxidation over Pd–Al$_2$O$_3$. A fractional factorial design was used for the experimental runs. Analysis of these data was completed in a subsequent article by Kittrell *et al.* (*41*). The rate expression (Hougen-Watson type) developed to fit the experimental data of Hunter and Mezaki adequately was

$$r = \frac{k \sqrt{K_2}\, p_{CH_4}\, p_{O_2}^2}{[1 + \sqrt{K_2}\, p_{O_2} + K_3\, p_{CO_2} + K_4\, p_{H_2O}]^3}$$

Kittrell *et al.* were careful to point out that . . . "no claim is made concerning the mechanism of the reaction or even the uniqueness of the model which has been set forth as adequately describing the experimental data." This caveat seems appropriate since one would have difficulty believing in the physical reality of the reaction mechanism which, by a straightforward Langmuir-Hinselwood approach for a third-order reaction between a methane molecule and two adsorbed oxygen molecules, would lead to the proposed rate equation.

Some dichotomy of thinking nevertheless appears to exist among the proponents of this approach. Box and Hill (*42*), in a paper entitled "Discrimination Among Mechanistic Models," propose to "discover the *mechanism* for a particular phenomenon leading to a specific mathematical model. . . . To discriminate among these [possible mechanisms] a sequential procedure is developed in which calculations made after each experiment determine the most discriminatory process conditions for use in the next experiment."

The proposed use of sophisticated computer analysis to establish or to disprove a mechanism has been the subject of considerable recent criticism. Boudart (1), for example, in commenting on a statistical reanalysis by Logan and Philip (43) of mechanistic data by Ozaki et al. on ammonia synthesis, observes "There is always a real danger in kinetics to treat data with a powerful method of analysis which may be far better than the data themselves." Allara and Edelson (44) have taken an even stronger position in a discussion of parameterization techniques in kinetics: "The ability to fit a set of numbers to a functional form resembling chemical kinetic equations cannot of itself establish the validity of the model. . . . We believe that it is most important to distinguish between the correlative value of parameter fits as opposed to the predictive capabilities of a truly fundamental model. . . . We especially object to the publication of these parameters as rate constants, a term which implies a fundamental property of the reaction, since this only further aggravates an already confused situation in the literature."

As a parenthetic comment, the author feels that a fundamental difficulty is not that statistical analysis is an inappropriately powerful tool for discriminating between a number of postulated models. The difficulty is in deciding whether any of the models has a sound theoretical basis.

Charles Ware (45) has called the author's attention to the application of a modified Box-Hunter approach to proceed from statistically designed, isothermal laboratory data to successful prediction of the performance and temperature distribution in adiabatic units. In brief, a power rate law suggested by literature information is used as a first approximation. The analysis of laboratory data indicates a possible need for modifying the form of the rate law, and iteration of the entire process finally results in a rate law satisfying the statistical criteria imposed. The empirical rate law is satisfactory for designing commercial units operating within the range of variables studied, and mechanistic conclusions are not drawn.

Hydrogenation of Propylene and Isobutylene. An exceptionally careful and fine study of olefin hydrogenation over Pt–Al_2O_3 was published by Rogers et al. (46) in 1966. The experimental kinetic data have been analyzed not only by the authors but also in at least three subsequent papers.

The final rate equation of Rogers et al. was:

$$r = \frac{\alpha K_1 K_2 p_1 p_2}{(1+K_1 p_1 + K_2 p_2)^2} + \frac{\beta K_1 K_2 p_1 p_2}{(1+K_1 p_1)(1+K_1 p_1 + K_2 p_2)}$$

(The subscript 1 refers to hydrogen, 2 to olefin.) This unusually complex form was arrived at rationally. First, the authors found that no single conventional (Hougen-Watson) equation fitted the experimental data.

They then resorted to a suggestion of Bond and Turkevich (47) (who had studied the deuteration of propylene over Pt–pumice in a static reactor) that there may be two simultaneous rate-controlling surface reactions between adsorbed hydrogen and adsorbed propylene. One is on sites where hydrogen and propylene are competitively adsorbed, the other on sites where hydrogen is adsorbed non-competitively. A Hougen-Watson formulation of this model (with undissociated hydrogen) leads to the rate equation shown; the first term involves hydrogen competitively adsorbed, the second non-competitively adsorbed. The terms α and β are the corresponding rate constants.

Excellent fits of the experimental data at three temperatures are obtained, and the parameters α, β, K_1, and K_2 show reasonable dependence on temperature (except for K_1 and K_2 for isobutylene, whose behavior is eccentric). The predicted curves for rate vs. olefin concentration show a maximum at low olefin concentration. All of the experimental points, by chance, fall on the descending branch of the calculated curve. [The question of maxima in rate curves is an exceedingly interesting one, which deserves discussion elsewhere. Since the experimental data of Rogers et al. did decrease monotonically with increasing olefin concentration, I examined the possible fit by a power rate law of the form $r = kp_1^a p_2^b$ (which cannot give a maximum). The greatest number of experimental runs was for propylene at 21°C. The four-parameter rate law of Rogers et al. was stated to fit the 21°C data points with an average deviation of 3.1%. The three-parameter rate law $r = 0.102\, p_1^{0.70} p_2^{-0.25}$ fits (data points read from curves) with an average deviation of 3.4%. No significance is attached to this result.]

One of the original authors, Lih, has recently re-interpreted the experimental data on isobutylene hydrogenation to involve a single rate constant, α, but also a fraction of the active sites, F_u, which is available for the chemisorption of hydrogen but not for that of isobutylene (48). Lih's modified Hougen-Watson type initial rate equation (still based on undissociated hydrogen) is of the form:

$$r = \frac{(1-F_u)\alpha K_1 K_2 p_1 p_2}{(1+K_1 p_1)(1+K_1 p_1 + K_2 p_2)} \left[1 - \frac{(1-F_u)K_2 p_2}{(1+K_1 p_1 + K_2 p_2)} \right]$$

The fit of the rate data is again good, as might be expected. The parameters α, K_1, and K_2 show reasonable variation with temperature; F_u increases with increasing temperature but apparently not in a simple manner.

Mezaki (49) revised the original equation of Rogers et al. to incorporate both dissociative adsorption of hydrogen (adsorption constant K_1) and adsorption without dissociation on single sites with no adjoining vacant sites (adsorption constant K_1'). The modified rate equation becomes:

$$r = \frac{\alpha\, K_1 K_2 p_1 p_2}{[1+(K_1 p_1)^{1/2} + K_2 p_2]^3} + \frac{\beta\, K_1' K_2 p_1 p_2}{(1+K_1' p_1)\,[1+(K_1 p_1)^{1/2}+K p_2]}$$

The behavior of the statistically fitted parameters is very reasonable for both propylene and isobuytlene. The equation is formidable, however; there are five adjustable parameters, each a function of temperature.

Finally, Kolboe (50) has objected to all the earlier treatments on the grounds that they ignored the fact, known from earlier mechanistic studies, that the hydrogen atoms of the olefin can undergo an isotopic exchange (with deuterium) parallel to the hydrogenation reaction. Limiting himself to only two sets of sites and to a dissociative (Farkas) model for the olefin adsorption, Kolboe is led to the following equation:

$$r = \sum_{j=1}^{2} \frac{k_j\,(K_{1,j}\,p_1)^{1/2}\,K_{2,j}\,p_2}{[1+K_{2,j}p_2/(K_{1,j}\,p_1)^{1/2} + (K_{1,j}\,p_1)^{1/2}]^3}$$

Again the fit to the original data of Rogers *et al.* is excellent, and the behavior of the parameters is plausible. However, there are now six adjustable parameters, each a function of temperature. [This proliferation of complications in olefin hydrogenation calls to mind a poignant remark by Rideal at the First International Congress on Catalysis (51): "Hinshelwood once called the hydrogen–oxygen reaction the Mona Lisa of chemical reactions. It may well be that her smile has been caught by the ethylene–hydrogen reaction at a nickel surface. A great number of workers in the field of catalysis from Sabatier onward have given explanations of the mechanism of the reaction, I myself have advanced three. At least two must be erroneous and, judging from the fact that no less than three communications are to be made on this subject during this week, it is quite likely that all three of them are wrong."]

Power Rate Laws. It may be appropriate to include here a few examples in which power rate law expressions of simple form have been successfully applied over a range of temperatures. Bohlbro (52) has published an extensive monograph on the kinetics of the water–gas shift reaction. He notes, as an example, that the Temkin-Pyzhev rate equation for ammonia synthesis may be deduced on the basis of the Langmuir, Temkin, or Freundlich isotherms, and that one cannot distinguish between isotherms by kinetic expressions (*see* discussion of this above). He concludes, in agreement with Kemball (53), that other techniques— isotopic species, infrared spectroscopy, adsorption measurements, etc.— are necessary to supplement kinetic experiments before drawing definite conclusions concerning reaction mechanisms in heterogeneous catalysis.

He then proceeds to analyze his own data on the shift reaction in terms of the rate expression

$$r = k(T)\ (CO)^l(H_2O)^m(CO_2)^n(1-\beta)$$

where

$$\beta = (CO_2)(H_2)/K(CO)(H_2O)$$

The values of l, m, and n deduced from experiments run at 330°–500°C are:

$T(°C)$	l	m	n
330	0.80	0.25	-0.65
360	(1.00)	—	—
380	1.00	0.20	-0.65
400	(1.00)	(0.20)	(-0.65)
420	1.00	0.25	-0.60
460	0.80	0.30	-0.60
500	0.80	0.35	-0.50

The interesting result, as Bohlbro notes, is the small variation of l, m, and n with temperature—i.e., the "apparent orders of reaction" with respect to CO, H_2O, and CO_2 are about the same at all temperatures. Bohlbro then states that a single rate expression,

$$r = k(T)\ \frac{(CO)^{0.90}\ (H_2O)^{0.25}}{(CO_2)^{0.60}}\ (1-\beta)$$

is valid with fairly good accuracy over the whole temperature range. The values of $k(T)$ deduced from this expression give an excellent Arrhenius plot ($\ln k$ vs. $1/T$), with an apparent activation energy of 27.4 kcal/mole.

An analogous result has been reported by Lunde and Kester (54) for the Sabatier reaction ($CO_2 + 4H_2 = CH_4 + 2H_2O$) over a Ru–$Al_2O_3$ catalyst. The results of 62 runs were fitted by a rate equation of the remarkably simple form:

$$-\frac{d(CO_2)}{dt} = k(T)\left\{(CO_2)^n(H_2)^{4n} - \frac{(CH_4)^n(H_2O)^{2n}}{K}\right\}$$

with $n = 0.225$ over the entire temperature range of 204° to 360°C. The apparent activation energy was 16.84 kcal/mole.

One of the most interesting applications of power rate laws to a complex reaction system is that reported by Hertwig et al. (55) for the selective oxidation of o-xylene to phthalic anhydride. Tolualdehyde was an analyzed intermediate; CO and CO_2 were undesired by-products.

After consideration of various reaction schemes, the following was arrived at as best representing the data:

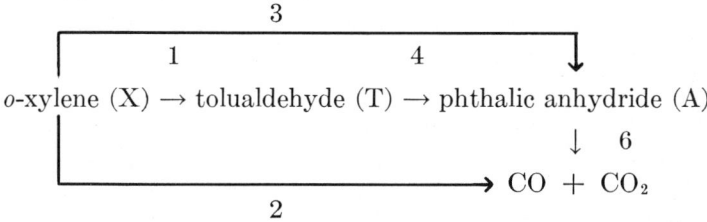

The individual rate equations were expressed as:

$$r_X = -(k_1 + k_2 + k_3)\, p_X{}^a$$
$$r_T = k_1\, p_X{}^a - k_4\, p_T{}^c$$
$$r_A = k_3\, p_X{}^a + k_4\, p_T{}^c - k_6\, p_A{}^d$$

The experimental data, obtained with a gradientless reactor, led to the following values of a, c, and d (the apparent orders of reaction) for three test temperatures:

$T(°C)$	a	c	d	Av. Error, %
360	0.46	0.40	0.50	11.7
380	0.41	0.39	0.50	10.1
400	0.40	0.39	0.52	25.4

The rate constants k_1, k_3, k_4, and k_6 showed reasonable Arrhenius behavior; k_2 showed no significant trend with temperature.

Again there is very small variation of a, b, and c with temperature. Hertwig et al. state that if average values are used, the equations

$$r_X = -(k_1 + k_2 + k_3)\, p_X{}^{0.45}$$
$$r_T = k_1\, p_X{}^{0.45} - k_4\, p_T{}^{0.40}$$
$$r_T = k_3\, p_X{}^{0.45} + k_4\, p_T{}^{0.40} - k_4\, p_A{}^{0.50}$$

can be used at all temperatures with an increase in the average error that is less than 2%.

There is an interesting supplement to the work just described. K. Lucas, who was responsible for this work, kindly informed the author (56) that despite the fit with a power function, the desire to design technical reactors with extrapolation to wide limits led the Leuna-Merseburg group to examine other rate law forms in which all the constants obeyed the Arrhenius expression. The results of the subsequent work have also been published (57). Equation sets of the Langmuir-Hinshelwood type were first examined but were discarded because some of the constants were negative; furthermore, six rate and seven adsorption constants would have been needed. The Mars-van Krevelen approach (*see below*) was then tried. A reaction scheme was successfully developed in which all the rate constants showed Arrhenius or approximate Arrhenius behavior. The rate equations are more complex in form than the power functions, and there are seven rate constants. Interestingly, the average errors in fitting the data were almost the same as with the earlier power rate laws: 12.1% at 360°, 10.1% at 380°, and 24.8% at 400°.

Mars-van Krevelen Models. In an important paper in 1952, Mars and van Krevelen (9) gave quantitative embodiment to a long used qualitative concept—*i.e.*, in oxidation reactions over transition metal oxide catalysts the reaction mechanism involves a cycling of the catalyst between lower and higher oxidation states of the metal ion. The Mars-van Krevelen approach has enjoyed considerable popularity; the recent paper of Mathur and Viswanath (58) contains many references to its use. My purpose here is only (1) to review the assumptions made in the experimental and theoretical work of Mars and van Krevelen, and (2) to mention that identical rate equations can be obtained by a steady-state Rideal-Eley mode. The latter point has already been clearly made by Downie and coworkers (59, 60), who prefer the expression "steady-state adsorption model," and it does not require further exposition here.

Mars and van Krevelen studied the oxidation of aromatic hydrocarbons in a fluidized bed of vanadium oxide catalyst. Since axial sampling of the vapor composition indicated some vertical gradient, they treated their fluidized bed reactor as an ideal plug flow reactor, with corresponding integration of the derived differential rate expression in order to treat the intergral conversions obtained. [That this cannot be correct is indicated, *inter alia*, by the fact that the measured naphthalene partial pressure near the inlet of the reactor, shown in their Figure 4 as *ca.* 0.035 mm, is far lower than even the lowest of the partial pressures (1 to 30 mm Hg) said to have been studied.]

A two-step reaction mechanism was postulated: (I) aromatic compound and oxidized catalyst → oxidation products and reduced catalyst. (II) Reduced catalyst and oxygen → oxidized catalyst. The reaction rates were assumed "to be of the first order with respect to the partial pressures of the substances to be oxidized." What this means for the catalyst is not specified, except in context. The rate of oxidation of the organic compound (Reaction I) is then taken as:

$$\dot{n}_R = k_1 \, p_R \, \theta$$

where R refers to organic reactant, and θ is "degree of coverage of the catalyst surface by the oxygen." The rate of re-oxidation of the surface reduced in the first reaction is assumed "to be proportional to a certain power of the partial pressure of the oxygen and to the catalyst surface not covered by oxygen." For Reaction II, therefore,

$$\dot{n}_{O_2} = k_2\, p_{O_2}{}^n\, (1-\theta)$$

If β moles of O_2 are needed for the oxidation of one mole of aromatic, $\dot{n}_{O_2} = \beta\, \dot{n}_R$. In the steady-state,

$$\beta\, k_1\, p_R\, \theta = k_2\, p_{O_2}{}^n\, (1-\theta)$$

which leads to the final rate expression

$$\dot{n}_R = \frac{k_1\, k_2\, p_R\, p_{O_2}{}^n}{\beta\, k_1\, p_R + k_2\, p_{O_2}{}^n}$$

The basis for choosing n is not clear. For aromatics oxidation, Mars and van Krevelen choose $n = 1$ apparently to permit easy integration of the rate equation. The integral equation gives reasonable fit to the experimental data although the authors point out that the activation energy for k_1 cannot be given "on account of the spread in the results."

Mars and van Krevelen use the same approach to fit the differential rate data of Krichevskaya (61) on SO_2 oxidation over vanadia but with the unexplained choice of $n = \frac{1}{2}$. [The published discussion following this paper is quite interesting. In response to a question from J. M. Smith concerning possible ambiguity in mechanisms, the authors reply: "Naturally, agreement between a kinetic formula and experimental data is not a proof of the correctness of the assumed reaction mechanism. In our investigation a mechanism made acceptable by experiments was taken as a basis for a consideration of the kinetics. The agreement between the kinetic formula and the experimental data makes it superfluous to consider other mechanisms and may be held to support the assumed mechanism."]

Conclusions and Suggestions

The preceding discussion probably makes clear the reasons for my own confusion. It seems difficult even to summarize the status of kinetic models in heterogeneous catalysis, hazardous to draw conclusions, and foolhardy to make recommendations. I offer here a personal evaluation which, to my regret, does not differ greatly from one I volunteered almost 20 years ago.

One first must decide his own needs and objectives. If these are to find a rate equation that is (a) consistent with all laboratory kinetic measurements and (b) useful for designing a reactor intended to operate within the limits of variables studied (temperature, partial pressures, conversion), then I know of no better suggestion than to use the statistically guided, iterative procedure described by Ware, deriving from the Box-Hunter approach. In this case I believe that the best model is the simplest one that works. The final rate equation may be a power rate law in form or Langmuir-Hinshelwood or any other that may prove to be more convenient. Within this context of practicality, a few further comments on the relative utility of the power and Langmuir-Hinshelwood forms may be useful.

A power rate law, where it applies, has a few merits. The experimentally determined powers immediately give a feeling for the way the rate depends on concentrations or partial pressures of reactants and products, in the same way that the apparent activation energy immediately gives a feeling for the sensitivity of the rate to temperature. The powers are, in fact, the apparent orders of reaction. A second point, also purely practical, is that a power rate law—where it applies—seems to give equivalent fit to experimental data with fewer adjustable parameters.

There are disadvantages to power rate laws. An obvious one is that there are many cases where the data cannot be adequately fitted by an equation of this form. A second one is the potential hazard in large extrapolations (although every rate law must really be treated very cautiously in extrapolating). Particularly troublesome in this regard are reactions, such as olefin hydrogenation or CO oxidation (62), in which the apparent order of reaction with respect to one reactant is found to be negative—*i.e.*, a reactant inhibits its own reaction. In the limit of sufficiently low concentrations, however, there must in principle be a range where the reverse behavior holds, and the rate increases with reactant concentration. This reversal, along with the implied rate maximum at some intermediate concentration, is not compatible with any single power rate law. By contrast, this behavior is totally compatible with the qualitative concepts of the Langmuir model for surface reactions, regardless of quantitative validity. In just such cases a Langmuir-Hinshelwood form may prove more useful for the rate law.

If one's objective is to get at the true mechanism of the reaction on a surface, it becomes much more difficult to make sensible suggestions because the problem is much more difficult. Probably the most important single caution is that measurement of global kinetics is singularly useless for determining catalytic mechanisms. This is demonstrable not only from theoretical considerations but also from published experience with a variety of reactions.

The determination of mechanism is always intriguing and sometimes feasible, but it should not be undertaken lightly. The few cases in which significant progress has been made typically involve the dedication of a professional lifetime by more than one researcher. The methodology for mechanistic research requires ingenuity in devising approaches uniquely relevant to the specific reaction of interest. A number of such approaches (not involving global kinetics) have been mentioned previously. A particularly powerful one, applicable only if the molecules undergoing reaction are sufficiently complex, is stereochemistry. In the hands of individuals such as Burwell, Pines, Kemball, Siegel, and others stereochemical characterization of product molecules has made it possible to draw strong conclusions about the mechanisms of catalyzed reactions of certain classes of organic compounds over specific catalysts. The mechanistic complexities of ammonia synthesis have certainly been illuminated by the careful application of isotopic tracer techniques, even though some fundamental uncertainties remain.

Suppose some information on mechanism is known independently through such studies. What approach should one take in developing a model for analyzing global kinetics? Unless quantitative information is available for the energy distribution in the specific catalyst–substrate system being investigated—a condition almost never fulfilled—I believe it a hopeless task rigorously to deduce a necessary rate law from any assumed model. Nevertheless, it is not unreasonable to fall back on the qualitative Langmuir concept that heterogeneous catalysis involves reactions between molecules chemisorbed in a monolayer. Existing mechanistic information (e.g., concerning rate-limiting steps) and assumptions concerning the possible form of adsorption isotherms may then be introduced to arrive at some form (model) for the rate law. Although I see no reason other than familiarity to prefer a Langmuir-Hinshelwood form over alternatives, neither do I see any reason not to start with such a form. If this is done, one is left with the usual requirement of demonstrating consistency (fit) with the experimental rate data. This leaves the normal Hougen-Watson fitting procedures intact, and I would sugegst only the following provisos:

(1) Models for which some "adsorption constant" is found to be negative should not be automatically discarded. Negative values may be expected if enhanced adsorption occurs.

(2) The human tendency to deduce mechanism through statistical data fitting should be resisted.

(3) If the independent mechanistic information leads to a rate equation that is inconveniently complex, there is no need to adhere rigidly to the form of the rate equation. I have encountered recent work in which mechanistic information was used to deduce a rate law, of the Langmuir-Hinshelwood form, containing eight adjustable parameters. This would not seem to be a profitable exercise.

(4) Since the rate-limiting step (if one exists) may change when temperature or concentrations are varied widely, one need not be surprised if the coresponding rate equations require similar change.

Acknowledgment

The author is most grateful to James Carberry and George Roberts for helpful discussions of some of the topics treated. They are not responsible for any of the opinions expressed.

Literature Cited

1. Boudart, M., *AIChE J.* (1972) **18**, 465.
2. Knözinger, H., Kochloefl, K., Meye, W., *J. Catalysis* (1973) **28**, 69.
3. Boudart, M., "Kinetics of Chemical Processes," p. 4, Prentice-Hall, Englewood Cliffs, N. J., 1968.
4. Langmuir, I., *J. Amer. Chem. Soc.* (1915) **37**, 1139.
5. Langmuir, I., *Trans. Faraday Soc.* (1921) **17**, 621.
6. Hinshelwood, C. N., "Kinetics of Chemical Change," p. 145, Clarendon Press, Oxford, 1926.
7. Rideal, E. K., *Proc. Cambridge Phil. Soc.* (1939) **35**, 130.
8. Eley, D. D., Rideal, E. K., *Proc. Roy. Soc.* (1941) **A178**, 429.
9. Mars, P., van Krevelen, *Chem. Eng. Sci. Spec. Suppl.* (1949) **3**, 41.
10. Sinfelt, J. H., *Advan. Catalysis* (1973) **23**, 91.
11. Langmuir, I., *J. Amer. Chem. Soc.* (1918) **40**, 1361.
12. Fowler, R. H., *Proc. Cambridge Phil. Soc.* (1935) **31**, 260; (1936) **32**, 144.
13. Langmuir, I., *J. Chem. Soc.*, **1940**, 511.
14. Boudart, M., *AIChE J.* (1956) **2**, 62.
15. Luckner, R. C., McConchie, G. E., Wills, G. B., *J. Catalysis* (1973) **28**, 63.
16. Hattikudur, V. R., Thodos, G., ADVAN. CHEM. SER. (1974) **133**, 80.
17. de Boer, J. H., *Advan. Catalysis* (1957) **9**, 472.
18. Thomas, J. M., Thomas, W. J., "Introduction to the Principles of Heterogeneous Catalysis," pp. 29-32, Academic Press, New York, 1967.
19. Weller, S. W., *AIChE J.* (1956) **2**, 59.
20. Gupta, R. B., Viswanathan, B., Sastri, M. V. C., *J. Catalysis* (1972) **26**, 212.
21. Gerai, S. V., Rozhkova, E. V., Gorokhovatsky, Ya. B., *J. Catalysis* (1973) **28**, 341.
22. Hertl, W., Farrauto, R. J., *J. Catalysis* (1973) **29**, 352.
23. Nakata, T., Matsushita, S., *J. Phys. Chem.* (1968) **72**, 458.
24. Stone, F. S., *Advan. Catalysis* (1962) **13**, 1.
25. Halsey, G. W., *Advan. Catalysis* (1952) **4**, 259.
26. Clark, A., "The Theory of Adsorption and Catalysis," pp. 46-57, Academic Press, New York, 1970.
27. Brunauer, S., Love, K. S., Keenan, R. G., *J. Amer. Chem. Soc.* (1942) **64**, 751.
28. Sips, R., *J. Chem. Phys.* (1950) **18**, 1024.
29. Brill, R., *J. Catalysis* (1970) **16**, 16.
30. Ozaki, A., Taylor, H. S., Boudart, M., *Proc. Roy. Soc.* (1960) **A258**, 47.
31. Brill, R., Richter, E. L., Ruch, E., *Angew. Chem., Intern. Ed. English* (1967) **6**, 882.
32. Schmidt, W. A., *Angew. Chem., Intern. Ed. English* (1968) **7**, 139.
33. Brill, R., Jiru, P., Schulz, G., *Z. Phys. Chem. N.F.* (1969) **64**, 215.
34. Hutchinson, H. L., Barrick, P. L., Brown, L. F., *AIChE Symp. Ser. No. 72* (1967) **63**, 18.
35. Kabel, R. L., Johanson, L. N., *AIChE J.* (1962) **8**, 621.
36. Knözinger, H., Ress, E., *Z. Phys. Chem. (Frankfurt)* (1967) **54**, 136.
37. Sridhar, T. S., Ruthven, D. M., *J. Catalysis* (1972) **24**, 153.
38. Benson, J. E., Kwan, T., *J. Phys. Chem.* (1956) **60**, 1601.
39. Box, G. E. P., Hunter, W. G., *Technometrics* (1962) **3**, 311.

40. Hunter, W. G., Mezaki, R., *AIChE J.* (1964) **10**, 315.
41. Kittrell, J. R., Hunter, W. G., Mezaki, R., *AIChE J.* (1964) **12**, 1014.
42. Box, G. E. P., Hill, W. J., *Technometrics* (1967) **9**, 57.
43. Logan, S. R., Philp, S. R., *J. Catalysis* (1968) **11**, 1.
44. Edelson, D., Allara, D. L., *AIChE J.* (1973) **19**, 638.
45. Ware, C., personal communication.
46. Rogers, G. B., Lih, M. M., Hougen, O. A., *AIChE J.* (1966) **12**, 369.
47. Bond, G. C., Turkevich, J., *Trans. Faraday Soc.* (1953) **49**, 281.
48. Lih, M. M., *J. Phys. Chem.* (1970) **74**, 2245.
49. Mezaki, R., *J. Catalysis* (1968) **10**, 238.
50. Kolboe, S., *J. Catalysis* (1972) **24**, 40.
51. Rideal, E. K., *Advan. Catalysis* (1957) **9**, 10.
52. Bohlbro, H., "An Investigation on the Kinetics of the Conversion of Carbon Monoxide with Water Vapour over Iron Oxide Based Catalysts," 2nd ed., Gjellerup, Copenhagen, 1969.
53. Kemball, C., *Act. IIeme Congr. Intern. Catalysis Paris, 1960*, **I**, 14.
54. Lunde, P. J., Kester, F. L., *J. Catalysis* (1973) **30**, 423.
55. Hertwig, K., Lucas, K., Flock, W., Bucka, H., *Chem. Techn.* (1971) **23**, 584.
56. Lucas, K., personal communication, 1972.
57. Hertwig, K., Lucas, K., Flock, W., Bucka, H., *Chem. Techn.* (1972) **24**, 393.
58. Mathur, B. C., Viswanath, *J. Catalysis* (1974) **32**, 1.
59. Shelstad, K. A., Downie, J., Graydon, W. F., *Can. J. Chem. Eng.* (1960) **38**, 102.
60. Juusola, J. A., Mann, R. F., Downie, J., *J. Catalysis* (1970) **17**, 106.
61. Krichevskaja, Je. L., *J. Phys. Chem. (USSR)* (1947) **21**, 287.
62. Carberry, J. J., personal communication.

RECEIVED December 4, 1974.

3

Contacting Effectiveness in Trickle Bed Reactors

CHARLES N. SATTERFIELD

Department of Chemical Engineering, Massachusetts Institute of Technology, Cambridge, Mass. 02139

> *The use of trickle bed reactors in industrial processing is reviewed, typical operating conditions are cited, and recent laboratory and pilot plant studies are summarized. Conversion in a trickle bed reactor usually comes closest to that for an ideal plug flow reactor at the highest liquid flow rates. Models to explain and to predict performance may be based on liquid holdup, catalyst wetting characteristics, liquid residence time distribution, or axial dispersion; these models are analyzed critically and compared. Some effects of transport limitations when reactant is present partially in the vapor phase are considered briefly.*

The central problem in design, scale-up, and understanding of the operation of trickle bed reactors is determining the effectiveness with which the reacting liquid is brought into contact with the solid catalyst. The liquid and gas flow rates used by the laboratories, pilot plants, and commercial operations vary greatly, and the wide range in liquid flow rate in particular is a primary cause of the variation in behavior with scale of operation. A brief review of industrial practices and the reasons therefore sets the stage for consideration of the effect of liquid flow rate on performance and for characterization of the effectiveness of catalyst contacting in a trickle bed reactor.

Industrial Processing

Petroleum Refining. The term trickle bed as used here means a reactor in which a liquid phase and a gas phase flow cocurrently downward through a fixed bed of catalyst particles while reaction takes place. These reactors have been used to a moderate extent in chemical processing, but most of the published information about their industrial applications concerns petroleum processing, in particular the hydrodesulfurization or hydrocracking of heavy or residual oil stocks and the hydrofinishing or hydrotreating of lubricating oils. According to van Deemter (1), a trickle bed hydrodesulfurization process was developed by Vlugter, Hoog, and their co-workers at the Shell Laboratories in Amsterdam after World War II, and the first commercial unit was brought on stream in 1955. The Shell process was described further by LeNobel and Choufoer (2). Similar developments occurred more or less simultaneously at other petroleum companies. Lister (3) described mixed-phase desulfurizer reactors that were developed by the British Petroleum Co. and put into operation during the 1950's; he gave considerable detail about various engineering design problems and how they were overcome. In the Proceedings of the various World Petroleum Congresses were published general descriptions of the hydrocracking and hydrodesulfurization processes developed by Chevron, Esso, Gulf, Union Oil, and other companies.

Trickle bed processing is less costly than completely vapor-phase operation since less heat is required for feedstock vaporization and less gas must be recycled and heated to reaction temperature. Moreover, trickle bed operation allows the processing of heavy distillates and residual oils that cannot react as vapors. Hydrocracking processes are usually preceded by a hydrodesulfurization reaction that reduces the content of organosulfur and organonitrogen compounds to a sufficiently low level so that they can be tolerated by the hydrocracking catalyst. In these processes, as in lubricating oil treatment, the reaction is between hydrogen and a petroleum stock, and high hydrogen pressures are used in order to obtain long catalyst life. The lubricating oil processing may be a so-called hydrofinishing that removes primarily organic sulfur and nitrogen compounds and improves color with little hydrogenation of the petroleum feedstock, or it may be a more severe hydrotreatment which causes not only sulfur and nitrogen removal but also ring hydrogenation and subsequent hydrocracking of one or more of the saturated rings (having started with a multiple-ring aromatic feedstock) to produce lubricating oils of high viscosity index. In some systems, e.g. some hydrodesulfurization reactions, much of the fluid present may be near or above the critical point and phase behavior is uncertain.

An alternate and closely related form of contacting is cocurrent gas–liquid upflow. It apparently has not been used industrially for the processing of petroleum fractions, but it has been applied in some chemical operations.

Chemical and Petrochemical Processing. Trickle bed reactors have also been used on a substantial scale for chemical and petrochemical processing, but information is fragmentary. In the Fischer–Tropsch synthesis of liquid fuels from mixtures of H_2 and CO, various means of removing the substantial heat of reaction have been tried including recycle of inert hot oil through the reactor bed as well as the use of hot-gas recycle and slurry-type reactors. Storch, Golumbic, and Anderson (4) treated in detail the chemistry and technology of the Fischer–Tropsch synthesis as it was developed in Germany before and during World War II, together with contributions from other sources. The hot-oil recycle version of this process was developed in the 1930s by Duftschmid and co-workers at I. G. Farbenindustrie; they relied on evaporative cooling of the hot oil to remove a portion of the heat by utilizing cocurrent upflow through the catalyst bed. A later version of the United States Bureau of Mines also operated by cocurrent upflow but with little or no evaporative cooling, and it was regarded as probably more efficient. It was reported that this form of contacting gives better temperature control than trickle-type operation, but the upflow version was shortly thereafter superseded by an ebulliating bed reactor because of difficulties with catalyst cementation in the fixed bed. The industrial scale Fischer–Tropsch processes used in Germany and elsewhere all involved completely vapor-phase operation, but the oil-recycle process was studied in Germany in converters as large as 50 ft³ and by the United States Bureau of Mines in a 50-bbl/day capacity plant (5, 6, 7).

A process for synthesis of butynediol ($HOCH_2C{\equiv}CCH_2OH$) from aqueous formaldehyde and acetylene uses trickle bed flow over a copper acetylide catalyst and recycle of the product stream for heat removal. One industrial reactor was reported to be 58.5 ft high and 4.9 ft in diameter (8). Other trickle bed studies of this reaction are given by Bill, a reported by Bondi (9).

Krönig (10) described a trickle bed process that is used in one or more commercial plants for selective hydrogenation of acetylene in order to remove it in the presence of butadiene in C_4 hydrocarbon streams. Operation at 10–20°C and 2–6 kg/cm² pressure allows liquid-phase processing which reportedly gives long catalyst life, unlike gas-phase processing in which polymers rapidly build up on the catalyst.

A patent by Porter (11) describes some interesting operating aspects of a trickle bed catalyst hydrogenator used to convert an alkyl anthraquinone to the hydroquinone form, which is one step in a cyclic process for manufacturing hydrogen peroxide. (Upon subsequent contact with oxygen, the hydroquinone yields the quinone plus hydrogen peroxide, and the quinone is then recycled.)

Recent Experimental Studies. In view of the widespread use of trickle bed reactors on a very large scale in the petroleum industry, it is surprising that so little was published about the development, design, and operation of this type of reactor. A useful general review of gas–liquid–particle operations by Østergaard (12) cites the literature up to about 1965–1966 on contacting between a gas and a liquid in fixed beds. Unfortunately for present purposes, most of this literature is more relevant to absorption systems than to chemical reactors. A small number of laboratory scale studies of specific chemical reactions in trickle-bed reactors were reported (see Table I which also includes a recent, representative, laboratory scale, petroleum processing study). A number of other laboratory or pilot plant scale studies of petroleum refining reactions were described by Henry and Gilbert (23), the most pertinent of which are summarized in Table II. In some trickle bed studies of chemical reactions, attention was directed primarily to catalytic behavior or chemical kinetics, and these investigations have been omitted unless some portion of the work focussed on physical effects.

Comparison with Slurry Reactors. The principal alternative to a fixed bed with two-phase flow, either upward or downward, is a slurry reactor or ebulliating bed in which the catalyst particles, which must be substantially smaller, are in motion. These are also sometimes termed three-phase fluidized bed reactors or suspended bed reactors. These have the following advantages: (a) a high heat capacity to provide good temperature control, (b) a potentially high rate of reaction per unit volume of reactor if the catalyst is highly active, (c) easy heat recovery, (d) adapability to either batch or flow processing, (e) a catalyst that is readily removed and replaced if its working life is relatively short, and (f) possible operation at catalyst effectiveness factors approaching unity which is of special importance if diffusion limitations cause rapid catalyst degradation or poorer selectivity. Their disadvantages are as follows: (a) the residence time distribution patterns are close to those of a CSTR which makes it difficult to obtain high degrees of conversion except by staging; (b) catalyst removal by filtration may pose problems with possible plugging difficulties with filters (and the costs of filtering systems may be a substantial portion of the capital investment); and (c) the high ratio of liquid to solid in a slurry reactor allows homogeneous side reactions to become more important, if any are possible.

In the trickle bed reactor, the catalyst bed is fixed, the flow pattern is much closer to plug flow, and the ratio of liquid to solid present is much smaller. If heat effects are substantial, they can be controlled by recycle of the liquid product stream although this may not be practical if a very high percent conversion is desired (as in hydrodesulfurization) or if the product is not relatively stable under reaction conditions. The

Table I. Recent

Reaction	Reference
Oxidation of SO_2 on wetted C	Hartman & Coughlin (13)
Hydrogenation of crotonaldehyde	Kenney & Sedricks (14, 15)
Isomerization of cyclopropane	Way (16, 17)
Hydrogenation of α-CH_3 styrene	Pelossof (18, 19) [b]
Hydrogenation of benzene	Satterfield & Özel (20)
Hydrogenation of α-CH_3 styrene	Germain, et al. (21) [a]
Hydrotreating	Mears (22) [c]

[a] For $\rho = 1.0$, 1 kg/m² sec = 0.1 cm/sec.
[b] Flow over a vertical string of spheres. L calculated for a bed of spheres touching in a square pattern.

Table II. Pertinent Laboratory Scale Petroleum

Reaction	Reference
Hydrocracking of a heavy gas oil	Henry & Gilbert (23)
Hydrodenitrogenation of various compounds and of a catalytically cracked light furnace oil [a]	Flinn, et al. (24)
Hydrodenitrogenation of a lube oil distillate	Gilbert & Kartzmark (25)
Hydrogenation of aromatics in a naphthenic lube oil distillate	Henry & Gilbert (23)

[a] There are no data points in original reference; highest flow rates were not recorded.

most common type of trickle bed processing is hydrogenation, and most of the subsequent discussion refers to this type of reaction. The other reactant may be essentially all in the liquid phase or in both the liquid and the gas phases, and the distribution of reactant and products between gas and liquid phases may vary with degree of conversion. In a few circumstances, as in some versions of the Fischer–Tropsch process, the liquid is inert and serves as a heat-transfer medium, and the reaction occurs between reactants in solution and the catalyst.

Industrial Operating Conditions. Trickle bed reactors in the petroleum industry may be operated under a wide variety of conditions that depend on the properties of the feedstock and the nature of the reaction. The less reactive fractions, which tend to be in the higher boiling range and more viscous at ambient temperatures, are typically processed at the lower liquid flow rates. Representative superficial liquid velocities, L, are 10–100 ft/hr (0.83–8.3 kg/m² sec for a density of 1) for lubricating oils, heavy gas oils, and residual fractions and 100–300 ft/hr (8.3–25

Laboratory Scale Studies

Superficial Liquid Flow Rate,[a] kg/m² sec	Gas Flow Rate kg × vJ³ m² sec	$Re_L = \dfrac{d_p L \rho}{\mu}$
0.0043–0.06	15.4	
0.038	0.47	
0.26–2.1	3–28	~0.04–8
1.4–8.7	—	
0.9–3.0	28[c]	16–55
0.08–1.6	0.14–3.3	
0.19–0.76	1.6–7.2	2–8

[c] G for H_2 plus benzene vapor.
[d] Countercurrent flow.
[e] At 100 atm pressure; all other studies at 1 atm.

Refining Studies Cited by Henry and Gilbert (23)

L, kg/m² sec	Conversion, %
0.07–0.5	61–20
lowest:[a] 0.07	99.8% [furnace oil reacted at 371°C (700°F)]
lowest:[a] 0.035	80% [quinoline reacted at 316°C (600°F)]
0.025–0.14[b]	97–70% (low temperature reaction)
0.025–0.06[b]	98.5–95% (high temperature reaction)
0.03–0.25[b]	80–30% (low pressure)
	95–40% (high pressure)

[b] Taking reactor height as 3 ft; true height was not published.

kg/m² sec) for naphtha fractions when calculated with the assumption that the feed is entirely in the liquid phase. For lighter fractions, this is not generally the case, and much of the feed is actually present as vapor. The hydrogen flow-to-liquid flow ratio is commonly expressed in terms of volume of H_2 [expressed in standard cubic feet (scf) at standard temperature and pressure (STP)] per barrel of feed processed (scf/bbl). The superficial gas flow rate, G, becomes about

$$G = \frac{(L)\,(\text{scf/bbl})}{5.6} \text{ cm/sec}$$

where L is likewise in cm/sec. Representative values are 2000–3000 scf/bbl for hydrodesulfurization of a heavy gas oil, 5000 scf/bbl for hydrodesulfurization of a heavy residue, and 5000–10,000 scf/bbl for a hydrocracker. For mild hydrogen treatment, hydrofinishing, the hydrogen-to-feedstock ratios may be considerably smaller.

Lister (3) summarized the characteristics of seven British Petroleum Co. desulfurizers that were designed between 1952 and 1962. Operating conditions range as follows: liquid hourly space velocity (volume of liquid fed each hour per volume of reactor), LHSV, 1.4–8.0 hr^{-1}; operating temperature, 690°–790°F (365°–420°C); pressure, 500–1000 psig; recycle rate, 1000–4000 scf/bbl; and single catalyst bed depth, 8–21 ft. One reactor consisted of three beds, each 10 ft 10 in. deep; a second unit consisted of three identical reactors in series, each 8 ft 9 in. deep. Reactor diameters ranged from 4 to 7 ft.

Present day units may be considerably larger, and multiple-bed reactors are frequently used. The quantity of catalyst is typically divided into one to five beds, each 10–20 ft deep; in multiple-bed reactors, hydrogen is injected between the beds for temperature control—so-called cold shot cooling. In multiple-bed reactors, the catalyst beds may be equal in depth. More commonly, they increase in depth as the reaction proceeds, and the quantity of hydrogen injected at each point is adjusted in order to achieve the desired axial temperature profile which is specified so as to limit the adiabatic temperature rise along each bed to some maximum [typically $\sim 28°C$ (50°F) or less]. The gas-to-liquid ratio thus increases with flow through successive beds, and the amount of gas injected for cooling can readily exceed that furnished initially. The quantity of H_2 furnished usually far exceeds that needed for stoichiometric reaction, and it is usually determined primarily by the requirements for temperature control and perhaps sometimes to help achieve better liquid distribution or to prolong catalyst life. The maximum height of a single catalyst bed is determined by the importance of achieving redistribution of liquid and gas after some limiting bed depth is traversed or by the crushing strength of the catalyst. In present practice, this maximum seems to be about 20–25 ft.

Representative operating conditions for petroleum refining processes are typically total pressures of 500–1500 psi (substantially higher in a few cases) and temperatures of 345°–425°C (650°–800°F). Catalyst particles are typically 1/8–1/32 in. (0.32–0.08 cm) in diameter.

Of considerable importance in analyzing data on trickle bed reactors is the fact that in pilot plants the 2–6-ft high reactors (1–1.5 in. in diameter) are typically operated at about the same LHSV as is used commercially. For a specified value of LHSV, the true liquid superficial velocity is thus proportional to reactor length. A large commercial reactor of recent design may have as much as 60–80 ft total depth of catalyst; data for this scale-up may have been obtained in a pilot unit at the same LHSV but therefore at 1/10–1/15 of the superficial liquid flow rate and at a correspondingly lower gas rate. Since the pilot unit and the commercial unit may operate under somewhat different hydrodynamic flow conditions, their contacting efficiencies may be significantly different. In most reports of laboratory studies of any type of reaction, the liquid and gas flow rates were much lower than those used commercially (*see* Tables I, II, and III). Some of the recently reported laboratory scale studies involved rapid exothermic reactions at low liquid flow rates; consequently, heat effects were especially significant. Catalyst pellets were only partially wetted, and reactants were relatively volatile so that reaction occurred in both liquid and vapor phases (*see* below). The behavior of these systems may be significantly different from those in which the reactant is exclusively in the liquid phase (to which subsequent discussion pertains).

Table III. Representative Limiting Flow Conditions for Petroleum Processing

Reactor	*Superficial Liquid Velocity*		*Superficial Gas Velocity*[a]
	ft/hr	*kg/m² sec*[c]	*kg/m² sec*[c]
Commercial	10	0.83	0.0132
	to		0.066
	300	25.0	0.395
			1.97
Pilot plant[b]	1	0.083	0.0013
	to		0.0066
	30	2.5	0.0395
			0.197

[a] Values of G were calculated for 1000 and 5000 scf H_2/bbl; it was assumed that all hydrocarbon is in the liquid phase.
[b] Length of pilot plant reactor was assumed to be 1/10 that of commercial reactor.
[c] 1 lb/hr ft² = 1.36 × 10⁻³ kg/m² sec.

Another precaution in interpretation stems from the fact that almost all of the published information on performance of industrial trickle bed reactors has dealt with petroleum refinery operations such as hydrodesulfurization. The wide spectrum of compounds present, with different reactivities, requires some arbitrariness in describing the intrinsic kinetics. For five fractions of a flashed petroleum distillate, Bondi (9) reported hydrodesulfurization differential reaction rates over a Co/Mo/Al_2O_3 catalyst that varied by as much as a factor of 4 at 40% conversion and by a factor of 7 or so at 80% conversion. If these are all treated as one compound in reactor analysis, the effect is to increase the apparent order of reaction. A mixture of several species with different reactivity, each exhibiting true first-order kinetics, appears to follow some higher order reaction over a wide range of conversion since the less reactive species persist longer than the more reactive ones. On the other hand, a group of species with similar reactivity can frequently be treated adequately when they are deemed to follow first-order kinetics over a limited range of conversion. The best procedure to follow varies with circumstances.

Models for Design and Analysis

The Ideal Trickle Bed Reactor. The analysis of trickle bed performance under ideal circustances and with the assumption of simple first-order kinetics provides a point of departure for analysis of real cases. We assume the following: (a) plug flow of liquid, *i.e.* no dispersion in the axial or radial direction; (b) no mass or heat transfer limitations between gas and liquid, between liquid and solid catalyst, or inside catalyst particles (the liquid saturated with gas at all times); (c) first-order isothermal, irreversible reaction with respect to liquid (gaseous reactant present in great excess); (d) catalyst pellets completely bathed with liquid; (e) the reactant completely in the liquid phase; and (f) no vaporization or condensation.

If we consider a differential volume element across the reactor and set the rate of reaction in that element equal to the disappearance of reactant as the liquid passes through the element, then:

$$Fc_{in}dx = rdV \tag{1}$$

where F = liquid flow rate in cm³/sec, c_{in} = concentration of reactant in entering liquid in moles/cm³, x = fractional conversion of reactant, dV = reactor volume in slice under consideration in cm³, and r = rate of reaction in moles/sec cm³ of reactor volume. If the reaction is first-order:

$$r = k_v c(1 - \epsilon) \tag{2}$$

where k_v = cm³ of liquid/cm³ of catalyst pellet volume sec. Substituting Equation 2 in Equation 1, we obtain

$$Fc_{in}dx = k_v(1 - \epsilon)cdv$$

Since $c = c_{in}(1 - x)$,

$$F \int \frac{dx}{(1-x)} = k_v(1 - \epsilon) \int dV$$

or

$$\ln \frac{c_{\text{in}}}{c_{\text{out}}} = \frac{V}{F} k_v (1 - \epsilon) = \frac{k_v (1 - \epsilon)}{L/h} = \frac{3600 k_v (1 - \epsilon)}{\text{LHSV}} \quad (3)$$

where V is the volume of the trickle bed packed with catalyst.

If the same simplifying assumptions again hold, we should be able to obtain the same values of the reaction rate constant k_v from studies in a stirred autoclave. In the autoclave, we measure change in concentration with time whereas in the trickle bed reactor the change is in concentration with distance. However, the autoclave and the trickle bed reactor should give the same value of k_v because there is a one-to-one correlation between time in the autoclave and distance traversed in the trickle bed. (For a specified flow rate, the distance traversed is inversely proportional to the dynamic liquid holdup, but it is unnecessary to know this in the ideal cease—*see* subsequent discussion on residence time distribution.)

In the autoclave:

$$-\frac{dc}{dt} = \frac{r(v_{\text{cat}} + v_{\text{liq}})}{v_{\text{liq}}} \quad (4)$$

where r is moles/(sec)(cm³ liquid + cm³ catalyst in autoclave), v_{cat} = volume of catalyst pellets in autoclave in cm³, v_{liq} = volume of liquid in autoclave in cm³, and t = time in sec. By substituting Equation 2 in Equation 4, where $(1 - \epsilon)$ is now the volume fraction of solid catalyst in the liquid slurry in the autoclave, we obtain:

$$-\frac{dc}{dt} \frac{v_{\text{liq}}}{v_{\text{cat}}} = k_v c$$

Integration gives:

$$\ln \frac{c_{\text{init}}}{c_{\text{final}}} = \frac{v_{\text{cat}} k_v t}{v_{\text{liq}}} \quad (5)$$

Although in principle k_v from Equation 5 should equal k_v from Equation 3, in practice that derived from Equation 5 is frequently greater than that calculated from trickle bed studies because of a loss of contacting effectiveness in the trickle bed.

Before this concept is discussed, however, other possible reasons for these differences should be noted. Homogeneous reactions, if they are possible, are more likely to be encountered in the autoclave than in the trickle bed because of the much higher ratio of liquid to catalyst volume. This could possibly lead to various unforeseen consequences such as more formation of side-product and polymers that might block pores. Catalyst poisoning would also probably cause the two reactor systems to behave differently. In the autoclave, poisoning would occur uniformly over all the catalyst particles whereas in the trickle bed poisons would usually be adsorbed preferentially onto the top-most catalyst layers leaving most of the bed clean for the main reaction. The net effect could well be a slower drop in reactivity with time in the trickle bed than would be expected from the autoclave studies. The distribution of poison through an individual catalyst particle may also be a function of particle size as well as of time and environment.

If the reaction is not actually first-order, the values of k_v as calculated from Equations 3 and 5 should be the same if the other assumptions hold and if comparison is being made for the same initial and final concentrations. In general this may not be the case, but, for many systems of interest where a high percent of conversion is not required, the kinetics of the reaction may be represented satisfactorily as a first-order process even when the true kinetics are substantially different.

Equation 3 is the same expression that is obtained for single-phase flow except that k_v is based on catalyst pellet volume and hence a factor $(1 - \epsilon)$ appears. Note that liquid holdup as such, or the true residence time of liquid in the reactor, does not appear in this expression. Neither does the aspect ratio of the reactor (ratio of length to diameter). The significance of these points is discussed below.

Contacting Effectiveness

At sufficiently low liquid and gas flow rates, the liquid trickles over the packing in essentially a laminar film or in rivulets, and the gas flows continuously through the voids in the bed. This is sometimes termed the gas continuous region or homogeneous flow, and it is the type that is usually encountered in laboratory and pilot scale operations. As gas and/or liquid flow rates are increased, one encounters behavior which is described as rippling, slugging, or pulsing flow, and this may be characteristic of the higher operating rates encountered in petroleum processing. At high liquid rates and sufficiently low gas rates, the liquid phase becomes continuous and the gas passes in the form of bubbles—this is sometimes termed dispersed bubble flow. This is characteristic of some chemical processing in which liquid rates are substantial, but the gas-to-liquid ratios are considerably below those encountered in much petroleum processing. Flow patterns and the transitions from one form to another as a function of gas and liquid flow rates were described by several authors and were recently summarized by Sato and co-workers (26).

If data are obtained over a range of linear velocities in a trickle bed, it is usually found that k_v as calculated from Equation 3 increases as the liquid flow rate is increased. In different terms, if both h and L are doubled (which keeps LHSV constant), the percent conversion is increased although Equation 3 predicts there should be no change. Furthermore, in the absence of complications as cited above, the value of k_v at the highest flow rates approaches the value obtained in a stirred autoclave where the catalyst is completely surrounded with liquid. Clearly, in most trickle bed reactors there is a loss in what is termed contacting effectiveness below that which can be obtained in the ideal reactor, and this loss is greatest at the lowest liquid flow rates.

In the past, designers of trickle bed reactors generally used the fact that contacting effectiveness improves with liquid flow rate as a built-in factor of safety by scaling-up from pilot plant to commercial plant size on the basis of equal values of LHSV. Since commercial reactors may be 5–10 times longer than pilot units, this would result in plant units operating at 5–10 times the liquid superficial velocity used in the pilot plants for the same value of LHSV. However, the situation was confused by the fact that in some cases, as reported by Ross (27), the commercial unit performed more poorly than the pilot plant unit in spite of the use of higher linear liquid velocities. This apparently was caused by poorer liquid distribution and was characterized by the finding that liquid holdup was poorer in the large unit (27). (This method of scale-up is unlikely to introduce difficulties from mass transfer limitations unless the catalyst particle size is increased at the same time which, however, the designer may be tempted to do in order to avoid excessive pressure drop.)

Let us now use the term apparent reaction rate constant, k_{app}, to refer to the value of the reaction rate constant as calculated from Equation 3 or the equivalent, applied to results from a real trickle bed, and retain the symbol k_v to refer to the intrinsic value. We shall now consider methods of predicting the effect of the liquid flow rate on k_{app}.

Bondi (9) developed an empirical relationship of such a form as to cause k_{app} to approach k_v as the superficial liquid velocity approaches infinity. His equation, expressed in terms of reaction rate constants, is:

$$\frac{1}{k_{app}} - \frac{1}{k_v} = \frac{A'}{L^b}$$

For a number of systems, $0.5 < b < 0.7$ with a median value of about 2/3. Bondi also reported that an increase in the gas rate moderately increased conversion, and he suggested multiplying L in his correlation by $(\rho_G G)^\beta$ where, for much of his data, $0.22 < \beta < 0.5$. However, in a reacting system such as this, changing the gas flow rate also changes the partial pressures of products and reactants such as H_2S and hydrogen. Since H_2S has some inhibiting effect on the reaction rate, the improvement associated with increased gas flow rate is probably partly chemical and partly physical. The usefulness of this approach is that a contacting effectiveness can be directly defined as k_{app}/k_v. In trickle phase studies of hydrodesulfurization of a heavy gas oil, this ratio was about 0.12–0.2 at $L = 0.08$ kg/m² sec and about 0.6 at 0.3 kg/m² sec. At low liquid flow rates, one would expect the ratio to be proportional to the 0.5–0.7 power of the liquid velocity, with the power decreasing at higher liquid velocities.

Liquid Holdup. Another approach derives from the fact that as liquid flow rate is increased, liquid holdup increases also. For the ideal reactor, the drop in conversion should follow Equation 3, but with increased holdup it is generally found that c_{out} is lower than predicted. The increased holdup is effective because it improves contacting or decreases axial dispersion or both. Regardless of the reason, it is remarkable that a variety of data can be brought into consonance by assuming (all other things being equal) that the rate of reaction is proportional to the holdup, H, or that

$$k_{app} \propto k_v H \qquad (6)$$

For a first-order reaction, one usually plots $\ln c_{out}/c_{in}$ against residence time in order to obtain the value of k, i.e.

$$\ln \frac{c_{out}}{c_{in}} \propto - k_{app} t \qquad (7)$$

If $k_{app} \propto k_v H$, then

$$\ln \frac{c_{out}}{c_{in}} \propto \frac{-k_v H}{(LHSV)} \qquad (8)$$

Ross (*27*), for example, showed that data on hydrodesulfurization for two commercial reactors and a pilot hydroreactor could be brought into agreement by plotting percent sulfur retention *vs.* (1/LHSV) · H on semilog coordinates. Here the commercial reactors operated more poorly even though higher liquid velocities were used. Henry and Gilbert (*23*) developed this concept further. For laminar flow down a bed of spheres, the holdup should be proportional to $Re^{1/3}$ which can be derived analytically and which was found experimentally by Pelossof (*18, 19*) for flow over a vertical string of spheres.

If one postulates (a) that reaction rate is proportional to holdup and (b) that holdup is proportional to $L^{1/3}$, then Equation 3 is replaced by

$$\ln \frac{c_{\text{out}}}{c_{\text{in}}} \propto \frac{-k_v h^{1/3}}{(\text{LHSV})^{2/3}} \qquad (9)$$

It is generally observed that the degree of conversion increases as the bed length is increased at a constant value of LHSV, *i.e.* if liquid velocity is increased in proportion to the increase in bed depth. This was ascribed to decreased axial dispersion (*see* below), but Equation 9 would predict the same effect which is attributed to increased holdup. Henry and Gilbert obtained good correlation of two sets of data on the effect of bed height on percent conversion by using Equation 9 (the flow rates used are not certain). They also demonstrated that, for a fixed bed length, Equation 9 was followed for several other sets of data (Table II) on hydrocracking, hydrodenitrogenation, hydrodesulfurization, and hydrogenation—all of various petroleum fractions (hydrodenitrogenation included some studies of model compounds). This remarkably good correlation was obtained in spite of two facts: (a) there is no theoretical justification for assuming that the rate of reaction is proportional to holdup and (b) as Henry and Gilbert noted, Equation 9 applies only when $H \propto L^{1/3}$. As the Reynolds number is increased, the dynamic regime changes from gravity–viscosity to gravity–inertia, and the exponent of the Reynolds number increases to a value substantially greater than 1/3.

The Henry and Gilbert correlations were presented in terms of LHSV values, and in order to compare them with other work, the range of liquid flow rates used in each of the most pertinent studies was estimated (Table II). When reactor height was not given, an arbitrary value of 3 ft was assumed. Although this does result in some error, rough values for liquid superficial velocity can then be calculated. The corresponding percent conversions are also cited; in each case, the correlating line went through 0% conversion at infinite flow rate. Certain reservations remain. At very high percent conversion, axial dispersion can be a significant factor, and it is still uncertain whether the spectrum of compounds present in all cases but one can be properly treated by simple first-order kinetics. Nevertheless, the Henry and Gilbert finding that the contacting effectiveness is proportional to the one-third power of the liquid flow rate seems to hold with flow rates up to at least ~ 0.5 kg/m² sec.

In contrast, Bondi suggested a somewhat higher proportionality in this flow region. The point at which catalyst contacting approaches 100% effectiveness remains elusive. In analyzing Bill's data on formation of butynediol from C_2H_2 and HCHO and Hofmann's data on hydrogenation of glucose to form sorbitol, Bondi noted that contacting effectiveness was still improving with flow rate at velocities as high as 3 kg/m² sec. In hydrodesulfurization studies, a modest further improvement in contacting effectiveness accompanied an increase in the liquid flow rate from about 7 to 10 kg/m² sec.

If reactivity were indeed proportional to holdup, for the laminar film model, $H \propto d_p^{-2/3}$ and to $\nu^{1/3}$ where $\nu = \mu/\rho$. Thus the Henry and Gilbert correlation becomes:

$$\ln \frac{c_{in}}{c_{out}} \propto h^{1/3}(\text{LHSV})^{-2/3} d_p^{-2/3} \nu^{1/3} \qquad (10)$$

This predicts that decreasing catalyst size will increase conversion, but the same general effect would be produced by varying catalyst size if diffusion limitations within catalyst particles were significant. Also, according to this model, a decrease in holdup caused by gas flow would decrease conversion, instead of the opposite as was found by Bondi. Furthermore, it seems unlikely that conversion would increase with an increase in viscosity.

Wetting Characteristics. Mears (*28*) questioned the Henry–Gilbert correlation and suggested that it is more realistic to assume instead that the reaction rate is proportional to the fraction of the outside catalyst surface which is effectively (freshly) wetted by the flowing liquid. Several investigators reported that the wetted area of packed beds at moderate liquid flow rates is proportional to the 0.25–0.4 power of the mass velocity, and, in a very recent correlation, Puranik and Vogelpohl (*29*) reported that the wetted area is proportional to the 0.32 power of the liquid velocity. By applying this correlation for ratio of wetted area to total area and by introducing the effectiveness factor, Mears obtained:

$$\log \frac{c_{in}}{c_{out}} \propto h^{0.32}(\text{LHSV})^{-0.68} d_p^{0.18} \nu^{-0.05} (\sigma_c/\sigma)^{0.21} \eta \qquad (11)$$

The term σ_c/σ relates to surface tension properties that are presumably constant for a given combination of liquid and packing, as are also η and d_p. Note that the form of Equation 11 with respect to h and LHSV is essentially the same as that of Equation 9.

Mears compared Equations 10 and 11 with available pilot plant data in various ways. Equation 11 holds only when the proportionality of Puranik and Vogelpohl is valid. When mass velocities are above about 1000 lb/hr ft² (about 1.5 kg/m² sec), a different procedure is recommended in order to allow for the asymptotic approach to complete wetting as the liquid flow rate is increased.

Little information has appeared on the effect of the gas flow rate other than the suggested correlation of Bondi which predicts a moderate increase in the apparent rate constant with increased gas flow. This would be expected to decrease holdup by the drag effect of the gas and to break up rivulets and to improve contacting. Although Henry and Gilbert did not determine the effect of gas flow rate in their correlation, it is worth noting that the decrease in holdup caused by gas flow would, according to their model, decrease conversion rather than increase it as was found by Bondi. This further reinforces the view that liquid holdup *per se* is not the fundamental factor but, rather, contacting of catalyst with liquid and minimization of axial dispersion.

The above approach utilizing wetting characteristics has the merit of relating contacting effectiveness to a parameter which can be physically visualized. A film flowing downward uniformly over all the particles in a catalyst bed is not necessarily the most stable configuration since surface tension will tend to reduce the total film area. Observations of a trickle bed during reaction indicate that, even with excellent initial distribution, the liquid may gather into rivulets which tend to maintain their position with time. In some portions of the bed, catalyst pellets are bathed continuously with flowing liquid while in other portions the catalyst pellets, although wetted, do not have a liquid film on the surface. It is the active or freshly contacted fraction of the packed bed that is of concern. Although a portion or all of the remainder may be wetted, the stagnant or noncontacted fraction contributes nothing to reaction. The analogous problem in packed bed absorbers was long studied to determine the fraction of packing which is effectively wetted. Hobler (*30*) compared in detail the correlations developed by various workers through the late 1950's. Recent treatments are referenced by Mears (*28*). Some differences between the two systems, as listed below, should be borne in mind, but these do not appear to offer any major difficulty.

(a) Catalyst pellets are almost always porous. Except during operation at low liquid flow rates combined with exothermic reaction, the entire bed eventually becomes wetted by capillarity, even with rivulet-like flow. Absorption packings are nonporous, and a portion of the bed is typically not wetted. However, the portion of the reactor bed wetted by capillarity does not contribute significantly to reaction unless the vapor pressure of the reactant is substantial.

(b) Although most of the data are for shapes that are characteristic of absorber packings such as Berl saddles and Raschig rings rather than for the spheres, pellets, and extrudates that are of concern in trickle bed reactors, recent correlations indicate about the same velocity dependence for these various shapes.

(c) Quantitative values for the fraction of the bed which is effectively wetted or freshly contacted with flowing liquid depend on the method used for their measurement. For example, the effective area for absorption of a gas into a liquid is the total area of that liquid present that is not effectively stagnant. With trickle bed reactors, we are concerned with the outside area of solid pellets that is effectively contacted with flowing liquid.

By tracer studies, Wijffels and co-workers (*31*) measured the fraction of the stagnant holdup in packed beds which is not washed out during trickle flow; they took this as a measure of the fraction of the bed which is effectively noncontacted during operation. Measurements were made with 1–9.5-mm glass beads at superficial liquid velocities of 1.0–7 kg/m^2 sec using water primarily although there were a few runs with cumene. The fraction of the bed freshly contacted with flowing liquid decreased with increasing particle size in the 1–5 mm range. A model based on minimizing the total energy of the system in a horizontal plane likewise predicted this effect of particle size. The model could not be adequately tested with 5- and 9.5-mm particles because the individual particles were not completely wetted. Studies with cumene were confined to 1- and 3-mm particles, and it would be interesting to test the model with larger particles, with organic systems which have lower surface tension than water, and with more readily wet solid surfaces.

In this as in other studies, it was found that prewetting the bed can cause behavior substantially different from that encountered when an initially dry bed is used. At the moderate liquid flow rates characteristic of laboratory scale trickle beds, the time required to reach steady state with an initially dry bed may be several hours (*15, 20*), but this may be reduced significantly by preflooding. Surrounding catalyst pellets with finer inert material seems to improve wetting (*15*) and to cause substantially improved conversion at constant liquid flow rate (*22*). Heat effects can also cause the rate of wetting upon start-up to be much slower than that encountered in the absence of reaction. A useful next step in developing these approaches would be to characterize better the contact angles and spreading characteristics of representative organic liquids on porous catalysts.

Residence Time Distribution. Another approach, which is not exclusive of those above, is to determine the extent to which reactor performance suffers by deviations from the ideal plug flow model, as characterized by the residence time distribution (RTD) of the liquid. If we assume first-order kinetics, and if RTD data are available in the form of an exit age distribution function $E(t)$ and the degree of conversion is known or specified, the two can be related by the expression:

$$\frac{c_{\text{out}}}{c_{\text{in}}} = \int_0^\infty e^{-k't} E(t)\, dt \tag{12}$$

Here $E(t)dt$ is the fraction of the exit stream that is present in the reactor for a residence time between t and $(t + dt)$. For a plug flow reactor, the expression is:

$$\frac{c_{\text{out}}}{c_{\text{in}}} = e^{-k't_p} \tag{13}$$

A reactor efficiency can then be defined as (t_p/t_m) 100 where:

$$t_m = \int_0^\infty t E(t)\, dt \tag{14}$$

and where t_p is the residence time as calculated from Equation 13 that would produce the same fractional conversion as that found experimentally.

Equations 12 and 13 treat the system as though it were homogeneous and thus $k' = k_v(1 - \epsilon)/H$ sec^{-1} where the holdup H is assumed to be constant with length and, strictly, need not be known. However, it can be calculated from the $E(t)$ function by the expression $H = t_m(F/V)$, and it provides a useful measure of catalyst contacting. For large commercial reactors, tracer measurements appear to be the only practicable way to determine holdup.

Reactor efficiency can also be defined equivalently as the ratio of the length of an ideal (plug flow) reactor to that of a real reactor required to produce the same specified percent conversion where the true value of k' can be calculated from Equation 12. An example of the use of this method to analyze trickle bed performance was given by Murphree et al. (32). For 90% conversion, they found that a small pilot reactor operated at 90% efficiency and that a commercial desulfurization reactor operated at 40–60% efficiency under conditions of poor operation and 70–80% efficiency under good operating conditions. The same method of analysis was also applied by Cecil et al. (33) in a study of the desulfurization of gas oil or residual fuel oil in a small pilot reactor at liquid rates of 35–1500 lb/hr ft^2 (about 0.05–2.2 kg/m^2 sec). The concept of reactor efficiency as used here may be subject to misinterpretation since it is very sensitive to the percent conversion chosen for comparison at high degrees of conversion. Thus, a moderate degree of deviation from plug flow behavior can be insignificant at 90% conversion but serious at 99% conversion.

This approach demonstrates the effect of RTD on reactor effectiveness, but it does not address itself directly to contacting *per se;* the portion of the bed that is not contacted effectively cannot be identified. The use of Equations 12 and 13 is rigorously correct if the reaction is strictly first-order and if there are no radial concentration gradients. As emphasized by Schwartz and Roberts (34), the desired RTD is that of the liquid external to the catalyst pores, but a tracer measurement used to determine the E(t) curve will usually include as well some contribution from the large internal holdup which exhibits itself in the form of increased tailing.

Axial Dispersion. As an alternative to the use of RTD, small deviations from plug flow can be described instead by the axial dispersion model which involves only one parameter, the axial dispersion coefficient, which is usually expressed as a Peclet number. The dispersion coefficient is obtained by assuming that all the mixing processes involved follow the same functional relationship as Fick's laws regardless of the actual mechanism, an assumption that becomes increasingly dubious with large degrees of deviation from plug flow behavior. RTD measurements in trickle beds are better fitted by a two-parameter cross-flow model such as that used by Hoogendoorn and Lips (35), Hochman and Effron (36), and others. This model assumes that the holdup consists of stagnant pockets and liquid in plug flow, and the two adjustable parameters are the fraction of the total liquid in plug flow and an exchange coefficient. Schwartz and Roberts (34) made a detailed comparison of the effect on predicted reactor performance of various RTD data correlated in terms of the dispersion model vs. the cross-flow or an equivalent model. If one takes the RTD data of Hochman and Effron, which are provided in both forms, for representative reactor cases corresponding to Re_L values of 8 and 66, the ratio of required catalyst volume calculated by the dispersion model to that calculated by the cross-flow model is 1.03–1.09 at 80% conversion and 1.11–1.22 at 90% conversion. This suggests that the axial dispersion model may be the more conservative in general. It is also adequate for initial estimates as to whether deviation from plug flow will be significant in any specific case.

Mears (22, 28) presents the following criterion for the minimum h/d_p ratio that is required in order to hold the reactor length to within 5% of that needed for plug flow.

$$\frac{h}{d_p} > \frac{20m}{Pe_L} \ln \frac{C_{in}}{C_{out}} \qquad (15)$$

where $Pe_L = d_p L / D_l$ and m is the order of reaction. Thus, the minimum reactor height increases with the order of reaction, and it is very sensitive to fractional conversion at high degrees of conversion.

Hochman and Effron (36) reported dispersion data for cocurrent methanol and nitrogen flow of 600–5000 lb/ft² hr (0.8–7 kg/m² sec) in a 6-in. column packed with 3/16-in. glass spheres. Peclet numbers for the liquid varied from about 0.15 at $Re_L = 4$ to about 0.40 at $Re_L = 70$ with considerable scatter. Apparently there were significant, but random, variations in dispersion from point to point and day to day. The effect of gas rate (up to 8.35 cm/sec) was small. These values of Pe_L are 1/3 to 1/6 those for single-phase liquid flow at the same Reynolds number and can be compared to $Pe = 2$ for fully developed single-phase turbulent flow in packed beds. Charpentier and co-workers (37, 38, 39) likewise reported Pe_L values for trickle flow down to an order of magnitude less than that for single-phase flow at the same liquid Reynolds number, but with much scatter. Equation 15 demonstrates that, for representative laboratory scale trickle bed reactors of the order of a foot or so in length, axial dispersion may cause a significant deviation from plug flow behavior when conversions are roughly 90% or more.

There are about 15 other reported studies of liquid-phase axial dispersion, but these generally involved countercurrent air–water systems, almost invariably with Raschig ring and occasionally with Berl saddle packing, and frequently with flow conditions outside the range of interest in trickle bed reactors. A summary listing was given by Michell and Furzer (40) who also presented a recommended correlation based on considerable work of their own. This involves a Reynolds number, Re′, based on the interstitial rather than on the superficial velocity which must therefore be combined with an expression for the dynamic holdup, H_t, to relate superficial and interstitial velocities. The relationships are:

$$Pe_L = (Re_L')^{0.70} Ga^{-0.32} \qquad (16)$$

$$H_t = (0.68) Re_L^{0.80} Ga^{-0.44} a d_p \qquad (17)$$

where Ga is the Galileo number, $d_p^3 g_c \rho^2 / \mu^2$, and the liquid velocity used in the definition of Re_L is the superficial velocity. These relationships represent a refinement over similar correlations developed by Otake and Kunigita (41) and Otake and Okada (42). Representative values of Pe_L for countercurrent air–water flow through 0.25-in. Raschig rings are 0.25 at $Re_L = 10$ and 0.5 at $Re_L = 100$ (43), values which are reasonably close to those reported by Hochman and Effron.

Peclet numbers for the gas phase, as reported by Hochman and Effron, were correlated by the expression:

$$Pe_G = 1.8 Re_G^{-0.7} 10^{-0.005 Re_L} \qquad (18)$$

for Re_G values of 11 and 22 and for a range of Re_L values from 5 to 80. This leads to Pe_G values one to two orders of magnitude less than those encountered in single-phase gas flow, which are attributed, as was suggested earlier by DeMaria and White (44), to the gas phase "seeing" agglomerates of particles covered with a bridge of liquid as an effectively larger particle. This explanation is supported by the facts that wet packing in the absence of flowing liquid gives a large increase in dispersion over dry packing and that Pe_G decreases with increased Re_L. However, gas-phase dispersion is not of concern ordinarily in trickle bed processing.

Mass and Heat Transfer Effects. Methods of estimating when significant concentration gradients exist between gas and liquid, between liquid and solid, or within the porous catalyst are treated elsewhere (45).

Internal diffusion limitations are commonly expressed in terms of the catalyst effectiveness factor, η, which is defined as the ratio of the observed rate of reaction to that which would be observed in the absence of any internal concentration or temperature gradients. Methods of estimating η have now been developed in great detail (46, 47). For a first-order reaction, internal diffusion will be insignificant if

$$\frac{(d_p/2)^2 r(1-\epsilon)}{D_{eff} \cdot c_s} < 1 \tag{19}$$

where c_s is the concentration of the key reactant in the liquid (usually the dissolved gas) at the solid–liquid interface. Unless the size of the diffusing molecules is comparable to that of the pores,

$$D_{eff} = \frac{D\theta}{\tau} \tag{20}$$

where θ is the catalyst void fraction and τ is the tortuosity factor which usually has a value of ~ 4 (extreme values for the usual catalyst structures are about 2–7). Effectiveness factors for commercial hydrodesulfurization reactors are typically 0.36–0.6 for catalyst particles 5–6 mm in diameter, which is a mild degree of diffusion limitation.

Consider now some effects that may occur if reactant is present in both the liquid and the vapor phases. The maximum steady-state temperature difference between the center and the outside of a catalyst pellet, ΔT, occurs when diffusion is sufficiently limiting so that reactant concentration in the pellet center approaches zero. Then (48)

$$\Delta T = \frac{(-\Delta H) D_{eff} c_s}{\lambda} \tag{21}$$

where λ is the thermal conductivity of the porous solid. Even with a highly exothermic reaction, it is unlikely that ΔT will exceed more than a few degrees if the pores remain filled with liquid, unless the gas is much more soluble than hydrogen in most liquids. An example illustrates this: take the enthalpy change on reaction $-\Delta H = 5 \times 10^4$ cal/mole and a hydrogen solubility of 10^{-4} g mole/cm^3, which would require elevated pressure in typical hydrocarbons. Typical values of the other terms are $D_{eff} = 2 \times 10^{-5}$ cm²/sec and $\lambda = 3 \times 10^{-4}$ cal/sec cm°K. ΔT is 0.33°C.

If pores are filled with vapor, however, temperature differences in the hundreds of degrees are quite possible because D_{eff} values for vapors are three to four orders of magnitude greater than those for solutes and gas-phase concentrations are not lowered by as large a factor. The key limiting component is then usually vaporized reactant rather than hydrogen. Representative conditions are as follows: $-\Delta H = 5 \times 10^4$ cal/mole (this is now per mole of vaporized reactant), $D_{eff} = 10^{-2}$ cm²/sec, $c_s = 3 \times 10^{-5}$ (representing vaporized reactant present in small mole fraction but superatmospheric total pressure), and λ as before. ΔT becomes 50°C. This situation will not develop, of course, if the reactant does not have an appreciable vapor pressure.

If a reaction is substantially diffusion-limited when pores are filled with liquid reactant, then circumstances that cause the pores to become filled instead with vaporized reactant can cause a marked increase in reaction rate which is associated with the marked increase in diffusivity. Indeed, this was found experimentally by Sedricks and Kenney (15) in a study of the hydrogenation of crotonaldehyde. Liquid-phase reaction could presumably switch to vapor-phase reaction at a critical value of the local liquid flow rate below which the heat evolved could no longer be carried away by the flowing liquid. This effect can interact with liquid flow to cause temperature instabilities in various ways. Germain and co-workers (21) described a cyclic and irregular behavior of a trickle bed in which α-methylstyrene was hydrogenated to cumene; this could be explained plausibly by postulating that completely wetted pellets had low reaction rates whereas partly wetted pellets grew hotter because of insufficient heat transfer to flowing liquid, which led to high activity and formation of polymeric by-products. These in turn reduced catalytic activity, which allowed the pellet to become cool and wetted again. Removal of polymer by solution in the flowing liquid allowed activity to be restored and the cycle to recommence. Quantitative analysis of these types of coupling effects might be very helpful in revealing possible causes of instabilities in trickle bed reactors in general.

Summary and Conclusions

Many factors of importance in the design and characterization of trickle bed reactors have not been covered in this review. These include a description of the hydrodynamic flow regions, characterization of pressure drop and liquid holdup, and mass transfer limitations that may exist between gas and liquid, between liquid and solid, or within the catalyst particles. Conditions in which reactant is distributed between the liquid and vapor phases can give rise to behavior characteristics which were only mentioned. Methods of characterizing the contacting effectiveness of trickle bed reactors, however, are central to predicting their performance and the recent developments reviewed here give promise of putting these methods of characterization on a much more fundamental basis than heretofore.

Acknowledgments

Discussions with and comments from many individuals, including Peter Kehoe, David W. Mears, George Roberts, and Thomas K. Sherwood, are greatly appreciated.

Nomenclature

Some symbols that were used only once are defined at point of use and are not included in this list.

c	= concentration, g mole/cm^3; c_s, concentration at liquid–solid interface
D	= molecular diffusivity, cm^2/sec; D_{eff}, effective diffusivity in porous catalyst (Equation 20)
D_l	= axial dispersion coefficient, cm^2/sec
d_p	= catalyst particle diameter, cm
F	= liquid flow rate, cm^3/sec (Equations 1 and 3)
G	= gas superficial flow rate, cm/sec at STP or kg/m^2 sec
Ga	= Galileo number, $d_p g_c \rho^2/\mu^2$ (Equations 16 and 17)
g_c	= conversion constant
H	= holdup, cm^3 liquid/cm^3 empty reactor volume; H_f, free draining or dynamic holdup
h	= height of reactor, cm, or depth of packed bed, cm
k	= first-order reaction rate constant; k_v, intrinsic first-order reaction rate constant per unit volume of catalyst pellet, cm^3 liquid/cm^3 catalyst pellet volume sec; k_{app}, apparent value as found experimentally; $k' = k_v(1-\epsilon)/H$, sec^{-1} (Equations 12 and 13); k_{app}/k_v = contacting effectiveness
L	= liquid superficial flow rate, cm/sec or kg/m^2 sec
LHSV	= liquid hourly space velocity, volume of liquid fed to reactor each hour per volume of reactor, = 3600 L/h hr^{-1}
m	= order of reaction
Pe$_L$	= Peclet number for liquid phase, $d_p L/D_l$
r	= rate of reaction, g mole/sec cm^3 reactor volume
Re$_L$	= Reynolds number for liquid flow = $d_p L \rho/\mu$; Re$_g$ for gas flow = $d_p G \rho/\mu$
T	= temperature, °C
t	= time, sec

V	= reactor volume, cm^3
v_{cat}	= volume of catalyst pellets in autoclave, cm^3
v_{liq}	= volume of liquid in autoclave, cm^3
x	= fractional conversion of reactant

Greek

ν	= kinematic viscosity = μ/ρ
ϵ	= void fraction in catalyst bed, same as fraction of reactor volume not occupied by catalyst particles; $(1 - \epsilon)$ = ratio of catalyst pellet volume to empty reactor volume
η	= effectiveness factor, ratio of actual rate of reaction in a porous catalyst to that which would occur if pellet interior were all exposed to reactants at the same concentration and temperature as that existing at the outside surface of the pellet
μ	= viscosity, poises
ρ	= density, g/cm^3
τ	= tortuosity factor (Equation 20)
θ	= void fraction in porous catalyst particle

Literature Cited

1. van Deemter, J. J., *Chem. Reaction Eng., 3rd Eur. Symp., 1964,* 215.
2. LeNobel, J. W., Choufoer, J. H., *Proc. 5th World Pet. Congr. Sect. III, Paper 18,* Fifth World Petroleum Congr., New York, 1959.
3. Lister, A., *Chem. Reaction Eng., 3rd Eur. Symp., 1964,* 225.
4. Storch, H. H., Golumbic, N., Anderson, R. B., "The Fischer-Tropsch and Related Syntheses," Wiley, New York, 1951.
5. Crowell, J. H., Benson, H. E., Field, J. H., Storch, H. H., *Ind. Eng. Chem.* (1950) **42,** 2376.
6. Benson, H. E., Field, J. H., Bienstock, D., Storch, H. H., *Ind. Eng. Chem.* (1954) **46,** 2278.
7. Kastens, M. L., Hirst, L. L., Dressler, R. G., *Ind. Eng. Chem.* (1952) **44,** 450.
8. Brusie, J. P., Hort, E. V., "Kirk–Othmer Encyclopedia of Chemical Technology," 2nd ed., Vol. 1, p. 609, Interscience, New York, 1967.
9. Bondi, A., *Chem. Tech.* (March 1971) 185.
10. Krönig, W., *6th World Pet. Congr., 1963, Sect. IV, Paper 7.*
11. Porter, D. H., FMC Corp., U.S. Patent **3,009,782** (1961).
12. Østergaard, K., *Adv. Chem. Eng.* (1968) **7,** 71.
13. Hartman, M., Coughlin, R. W., *Chem. Eng. Sci.* (1972) **27,** 867.
14. Kenney, C. N., Sedricks, W., *Chem. Eng. Sci.* (1972) **27,** 2029.
15. Sedricks, W., Kenney, C. N., *Chem. Eng. Sci.* (1973) **28,** 559.
16. Way, P. F., Ph.D. Dissertation, Massachusetts Institute of Technology, 1971.
17. Satterfield, C. N., Way, P. F., *AIChE J.* (1972) **18,** 305.
18. Pelossof, A. A., Ph.D. Dissertation, Massachusetts Institute of Technology, 1967.
19. Satterfield, C. N., Pelossof, A. A., Sherwood, T. K., *AIChE J.* (1969) **15,** 226.
20. Satterfield, C. N., Özel, F., *AIChE J.* (1973) **19,** 1259.
21. Germain, A. H., LeFebvre, A. G., L'Homme, G. A., ADV. CHEM. SER. (1974) **133,** 164.
22. Mears, D., *Chem. Eng. Sci.* (1971) **26,** 1361.

23. Henry, H. C., Gilbert, J. B., *Ind. Eng. Chem. Process Des. Dev.* (1973) **12**, 328.
24. Flinn, R. A., et al., *Hydrocarbon Process Pet. Refiner* (1963) **42** (9), 129.
25. Gilbert, J. B., Kartzmark, R., *Proc. Am. Inst. Pet.* (1965) **45** (3), 29.
26. Sato, Y., Hirose, T., Takahashi, F., Toda, M., Hashiguchi, Y., *J. Chem. Eng. Jpn.* (1973) **6**, 315.
27. Ross, L. D., *Chem. Eng. Prog.* (1965) **61** (10), 77.
28. Mears, D. E., ADV. CHEM. SER. (1974) **133**, 218.
29. Puranik, S. S., Vogelpohl, A., *Chem. Eng. Sci.* (1974) **29**, 501.
30. Hobler, T., "Mass Transfer and Absorbers," Engl. transl., pp. 219–230, Pergamon, London, 1966.
31. Wijffels, J.-B., Verloop, J., Zuiderweg, F. J., ADV. CHEM. SER. (1974) **133**, 151.
32. Murphree, E. V., Voorhies, A., Mayer, F. X., *Ind. Eng. Chem. Process Des. Dev.* (1964) **3**, 381.
33. Cecil, R. R., Mayer, F. X., Cart, Jr., E. N., unpublished data.
34. Schwartz, J. G., Roberts, G. W., *Ind. Eng. Chem. Process Des. Dev.* (1973) **12**, 262.
35. Hoogendoorn, C. J., Lips, J., *Can. J. Chem. Eng.* (1965) **43**, 125.
36. Hochman, J. M., Effron, E., *Ind. Eng. Chem. Fundam.* (1969) **8**, 63.
37. Charpentier, J. C., Prost, C., LeGoff, P., *Chem. Eng. Sci.* (1969) **24**, 1777.
38. Bakos, M., Charpentier, J. C., *Chem. Eng. Sci.* (1970) **25**, 1822.
39. Charpentier, J. C., Prost, C., LeGoff, P., *Chim. Ind. Genie Chim.* (1968) **100**, 653.
40. Michell, R. W., Furzer, I. A., *Chem. Eng. J.* (1972) **4**, 53.
41. Otake, T., Kunigita, E., *Kagaku Kogaku* (1958) **22**, 144.
42. Otake, T., Okada, K., *Kagaku Kogaku* (1953) **17**, 176.
43. Furzer, I. A., Michell, R. W., *AIChE J.* (1970) **16**, 380.
44. De Maria, F., White, R. R., *AIChE J.* (1960) **6**, 473.
45. Satterfield, C. N., *AIChE J.*, (1975) **21**, 209.
46. Petersen, E. E., "Chemical Reaction Analysis," Prentice-Hall, Englewood Cliffs, N.J., 1965.
47. Satterfield, C. N., "Mass Transfer in Heterogeneous Catalysis," paperback edition, Dept of Chemical Engineering, Massachusetts Institute of Technology, Cambridge, 1970.
48. Prater, C. D., *Chem. Eng. Sci.* (1958) **8**, 284.

RECEIVED December 4, 1974. Work supported by the National Science Foundation.

4

Physical Processes in Chemical Reactor Engineering

E. WICKE

Institut für Physikalische Chemie, Westfälische Wilhelms-Universität, 4400 Münster, Germany

> *The scope of this review is the presentation and critical examination of the physical principles that underlie the pseudo-homogeneous models for processes in multiphase systems. For diffusion systems, i.e. porous gas–solid and SLP catalysts, this principle is effective diffusivity; its advantages, its shortcomings, and the limits of its applicability are discussed in detail. For heterogeneous flow systems, this principle is radial and axial dispersion. The elementary processes including the Taylor dispersion are illustrated; applications to gas phase and to solids dispersion in fluidized beds as well as to interphase heat transfer in packed beds are presented. The limits of pseudo-homogeneous modeling are indicated with the mass transfer problems of catalyst screens and of shallow beds. Differentiation between axial dispersion of mass in packed beds and real backmixing is emphasized. This difference requires a change of the well known boundary conditions of Dankwerts, and of Wehner and Wilhelm, at the bed entrance.*

Although the papers in the session on "Physical Processes" (*see* ADVANCES IN CHEMISTRY SERIES No. 133) are different in scope and nature, they contain a number of common viewpoints that may serve as a basis for this review. One of the most important viewpoints seems to be the obvious tendency to describe processes that run in heterogeneous multiphase systems by pseudo-homogeneous models and to comprehend the behavior of such processes, stationary or dynamic, in terms of single-phase concepts. Hence the theme of this review will be to indicate the principles and the methods of developing homogeneous models for heterogeneous processes and to present the usefulness of those concepts as well as the limits of their applicability.

There are two principal types of heterogeneous systems in chemical reactors (Table I): the flow systems where hydrodynamic flow is the predominant process of mass transport, and the diffusion systems where

Table I. Heterogeneous Reaction Systems

Type of System	Two Phases	Three Phases
Flow	packed beds fluidized beds	trickle beds bubble columns (with suspended catalysts)
	macroheterogeneity: dense emulsion } phase–bubble phase	
	microheterogeneity:	grains–fluid
	grids, screens, gauzes	
Diffusion	porous solids	SLP catalysts
	macropore and micropore systems	

molecular diffusion prevails. Typical flow systems are represented by packed beds, fluidized beds, grids, and screens, and those with three phases by trickle beds and bubble columns with suspended catalysts. Diffusion systems are the porous gas–solid catalysts and the supported liquid phase (SLP) catalysts. In some cases, different levels of heterogeneity can be discerned. In fluidized beds and in three-phase bubble columns there are the macro level of heterogeneity (dense or emulsion phase and bubble phase) and the micro level (grains and fluid); in diffusion systems, these levels of heterogeneity are the macropore and the micropore arrays.

In flow systems, the heterogeneity leads to the well known dispersion effects. Parallel to the main flow direction—axial or longitudinal—the dispersion is appreciably broader than it is normal to it—radial or transversal. In order to describe these effects in analogy with diffusion processes, meaningful coefficients for radial and axial dispersion may be measured and defined only if the system is large enough to cover the length of at least 10 characteristic dispersion pathways in either direction. If this condition is not satisfied, the model of the ideal single mixing stage or the cascade of mixing cells may be applicable to the system. If this fails also, then a pseudo-homogeneous model does not seem to be adequate for the flow system. This is true, for instance, for screen catalysts and for shallow packed beds as well as for certain fluidized beds (*see* below).

Diffusion Systems

The pseudo-homogeneous model can be applied in heterogeneous diffusion systems if the units of heterogeneity—pores, holes, particles, etc.—are small compared with the length of the diffusion paths, *i.e.* the extension of the concentration profiles. The most simple application of this model is to make use of an effective diffusion coefficient:

$$D_{\text{eff}} = \psi \cdot D_{\text{m}} \qquad (1a)$$

where ψ is the permeability of the porous body and D_{m} is the diffusion coefficient in the free gas phase. In order to measure D_{eff} accordingly, steady state diffusion must be applied (the term permeability indicates steady state conditions), and the gas pressure must be sufficiently high in order to justify the neglect of Knudsen diffusion. The permeability can then be used to define a labyrinth factor χ (or tortuosity factor $1/\chi$) by: $\psi = \epsilon_p \cdot \chi$ where ϵ_p is the porosity. The system of transport pores for permeation through the porous body is normally the same as the macropore system and has a fairly well defined mean pore radius, r; the micropores are then similar to pockets in the walls of the macropores. In such a case, the permeability in the Knudsen region, according to

$$D^K_{\text{eff}} = \psi^K \cdot D^K(r) \qquad (1b)$$

with $D^K(r) =$ Knudsen diffusion coefficient, can be considered equal to the permeability for bulk diffusion: $\psi^K \approx \psi$. If on the other hand, the micropores contribute markedly to the diffusion resistance for permeation, $\psi^K \neq \psi$ holds, which means that the permeability in the transition region becomes dependent on total gas pressure and on the type of molecular species that is diffusing.

For nonsteady state diffusion in porous systems, the mass balance equation is

$$\frac{\partial}{\partial t}(\epsilon_p c + n_a) = \text{div}(\psi D_{\text{m}} \cdot \text{grad } c) \qquad (2)$$

where n_a is the amount of gas adsorbed at the pore walls per unit volume of the porous body. Pseudo-homogeneous treatment is possible if n_a can be taken as the value in adsorption equilibrium with c. In the region of linear adsorption isotherm and with D_{m} independent of concentration, Equation 2 then simplifies to

$$\frac{\partial c}{\partial t} = \frac{\psi D_{\text{m}}}{\epsilon_p + dn_a/dc} \text{ div grad } c = D'_{\text{eff}} \cdot \text{div grad } c \qquad (3)$$

with an effective coefficient D'_{eff} for nonsteady state diffusion. When adsorption is negligible

$$D'_{\text{eff}} = \frac{\psi}{\epsilon_p} D_m = \frac{D_{\text{eff}}}{\epsilon_p} = \chi D_m \qquad (4)$$

rather than Equation 1a for steady state conditions. If, on the other hand, adsorption with curved isotherms has to be considered, the mass balance in the general form of Equation 2 must be used, and the solution of diffusion problems becomes very complicated although still within the concept of pseudo-homogeneity.

Wakao's paper (1) starts with the question if the binary counter-diffusion of inert, nonadsorbable gases through porous media runs for steady state and for nonsteady state conditions with effective diffusion coefficients that differ only by the factor ϵ_p, i.e. $D_{\text{eff}} = \epsilon_p \cdot D'_{\text{eff}}$ according to Equation 4. In fact, the validity of this relation could be confirmed by measurements with N_2–H_2 mixtures in porous refractory material as well as by calculations of the diffusion through a two-dimensional network of interconnected macro- and micropores. This network model was also used by Wakao to discuss the basic question if the effective diffusion coefficient, as measured by methods of steady state permeation through the porous medium, D^D_{eff}—binary bulk diffusion or Knudsen diffusion—will be the same as the coefficient, D^R_{eff}, that can be obtained from the effectiveness factor of a first-order reaction running at the pore walls. In fact, the calculation yields differences for the same molecular species with and without reaction: $D^R_{\text{eff}} < D^D_{\text{eff}}$. The reason is obvious: the micropores contain the predominant part of the internal surface area, and they are therefore rather strongly engaged in diffusional resistance when the transport runs to the points of reaction, compared with the case of permeation through the porous body without reaction. The type of model chosen, however, enhances this effect; in practice it will be rather small, as Wakao concludes.

Large differences between the values of steady state and unsteady state diffusion coefficients occur with SLP catalysts, where the catalytically active material is distributed in a liquid that covers the pore walls like a film and fills up the smallest (micro) pores. As long as the gas-phase diffusion resistance in the residual open pore cross-sections is determining for the transport rate, the system can be treated like porous gas–solid systems. Steady state diffusion can then be described by a permeability, Equation 1a, which in this case is a function of the amount of liquid loading, as Abed and Rinker (2) demonstrated. Nonsteady state diffusion of insoluble species leads to $D'_{\text{eff}} = D_{\text{eff}}/\epsilon_p$ where ϵ_p also depends on liquid loading. For nonsteady state processes with soluble species, the gas-phase diffusion is rate-determining only if the liquid is so thinly spread out along the pore walls that absorption equilibrium is established more quickly than the gas-phase concentrations of the species in the pore cross-section can change. If this is the case, Equation 2 can be applied with $n_a =$ amount of diffusion species absorbed in equilibrium. In the region of linear absorption, the simplified form of Equation 3 is applicable, wherein the slope of the absorption isotherm (dn_a/dc) represents a strong individual effect of the diffusion species and gives rise to appreciable differences between steady state and unsteady state diffusion. The reverse situation, in which diffusion resistance in the liquid phase is rate-determining in nonsteady state processes, is characteristic for gas–liquid chromatography; here the diffusion paths in the gas phase are kept short by using very small particles with coarse porosity.

A particular case of considerable importance in industry is the SLP catalyst for SO_2 oxidation, V_2O_5/K_2O melts on porous supports. Villadsen and co-workers (3) concluded that, in typical industrial catalysts for this reaction (liquid loading about 50% of the pore volume), the diffusion resistance is almost evenly divided between liquid-phase diffusion and pore diffusion. Even in this complicated case, the authors succeeded in developing a pseudo-homogeneous model. The liquid loading is thought to be composed of disc-shaped segments that cover the pore walls; the diffusion of the reaction components in these segments is supposed to be restricted to one dimension, *i.e.* normal to the front face. [Similar models were developed earlier in gas–liquid chromatography— see Giddings (4)]. The contribution of each segment to the local reaction rate at the pore wall is calculated by applying the effectiveness factor for plate geometry; an integration by means of a suitably chosen distribution function of segment thicknesses yields the local reaction rate per unit volume of the porous support, v_{eff}. This effective reaction rate is then used in the balance equations for pore diffusion in the residual pore volume:

$$\text{div}\,(D_{i,\,eff}\,\text{grad}\,c_i) = -\nu_i v_{eff} \qquad (5)$$

where $\nu_i =$ stoichiometric number and $\nu_i v_{eff} =$ rate of production of component i per unit volume of support. The decisive point for the applicability of this pseudo-homogeneous treatment is the tacit assumption that the absorption equilibrium is locally established at the gas–liquid phase boundaries. If, instead of equilibrium, a mass transfer resistance has to be taken into account at the phase boundary, then a pseudo-homogeneous model, *i.e.* treatment of each component by a single transport equation like Equation 5, is no longer applicable. Two transport equations, one for each phase, are then necessary; they are coupled by the interphase mass transfer relation.

The calculations of Villadsen and co-workers were simplified by the fact that in the liquid phase only the diffusion of oxygen had to be considered and that the rate of SO_2 oxidation could be taken as proportional to the distance of oxygen concentration from equilibrium (first-order). If, in more general cases, the diffusion resistances of more than one component in the liquid phase—and possibly multicomponent diffusion in the gas phase—have to be taken into account, the problem becomes much more complicated. Then the system of balance equations can no longer be reduced to one equation for a single component (Equation 5). Problems like these have not yet been solved; successful treatment should be possible by an extension of the methods of stoichiometric balances which were developed recently for multicomponent bulk diffusion of reaction components in porous media [Hugo (5, 6, 7), Hesse (8, 9)].

Dispersion Effects in Packed Beds

A packed bed of equally sized, spherical, nonporous particles is the exemplary case for investigating flow dispersion effects. The elementary processes to be considered here are presented in Figure 1. These are: the molecular diffusion in the voids between the particles (a), the branching effect of the solid packing that generates a network of interwoven streamlines that may be imagined as a random walk model (b) or as a capillary model (c), the eddy diffusivity (d) where the voids act as mixing cells, and finally the channeling (e) brought about by irregularities in the packing and by wall effects. The combined action of these dissipative processes in the direction transversal to the main flow can be described by a radial mixing coefficient:

$$M_r = \chi D_m + 0.1 \cdot u d_p \qquad (6)$$

The molecular diffusion D_m must be corrected with the labyrinth factor χ (see Equation 4). In a regular packing arrangement, $\chi \approx 0.7$. The radial dispersion, $D_r = 0.1 \cdot u d_p$ where u = mean axial flow velocity between the pellets and d_p = pellet diameter, is brought about by lateral displacement of the streamlines in the branching processes. With a simple geometry, the random walk model (Figure 1b) yields $1/8 \cdot u d_p$; measurements with gas and with liquid flow give a mean value of $1/10 \cdot u d_p$ [Wilhelm (10), Hiby (11)].

Parallel to the main flow direction, the dissipative processes give rise to a residence time distribution, RTD, which can be measured by

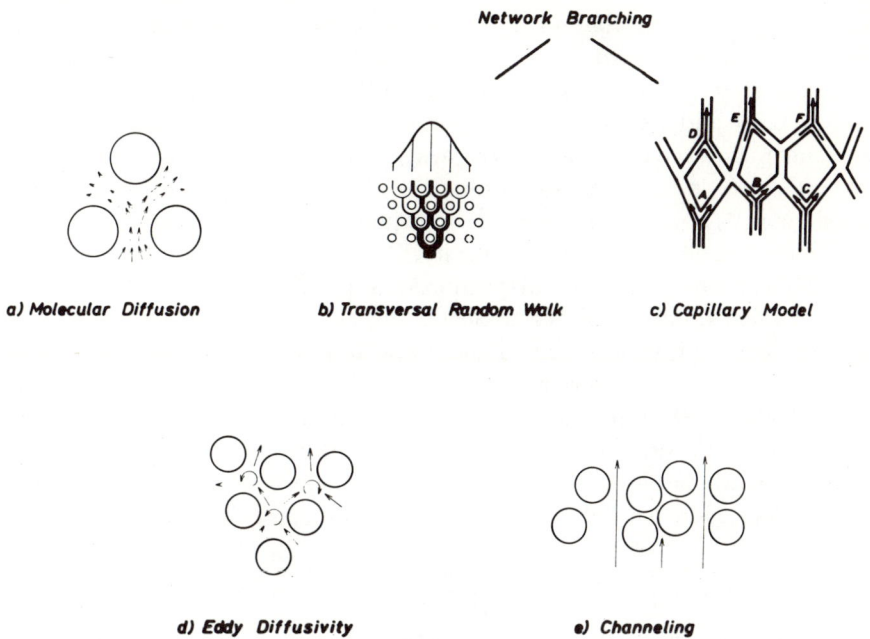

Figure 1. Packed bed diffusion and dispersion processes

tracer methods. The spreading-out effect can be described by an axial distribution coefficient:

$$M_{\text{RTD}} = \chi D_m + 0.5 \cdot u d_p \tag{7}$$

that includes the contributions of molecular diffusion, the same as in Equation 6, and of axial dispersion: $D_a = 0.5 \cdot u d_p$ (contributions from channeling and from wall effects are not taken into account). The axial dispersion comes from the retardations and accelerations of the flow between the pellets. Taking the pellet diameter as the plausible random walk distance, one obtains $D_a = 0.5 \cdot u d_p$ [Wicke (*12, 13*)]. As is well known, the expression can also be derived as an effect of eddy diffusivity by imagining a cascade of mixing cells (Figure 1d) as was done by Aris and Amundson (*14*).

The dispersion coefficients D_r and D_a have been defined in order to describe the dissipative processes in the heterogeneous packed bed by a pseudo-homogeneous formalism. This provides, of course, that the packed bed is sufficiently extended, at least over 10 pellet diameters in the radial and axial directions. The different values of D_r and D_a originate from the anisotropic nature of tubular flow and indicate the different physical meanings of these dispersion processes. D_r describes the radial flattening of concentration profiles as a consequence of random jumps back and forth with reference to control planes fixed in laboratory space, whereas D_a describes the axial flattening of concentration profiles by random jumps back and forth with reference to a control plane that moves with the mean flow velocity u, i.e. D_a describes the broadening of an RTD. In order to indicate these differences, the notation axial distribution coefficient M_{RTD} was introduced in Equation 7 instead of axial mixing coefficient or effective axial diffusivity.

Measurements of axial dispersion with gas flow through packed beds at Reynolds numbers above about 10 agree with the expectation from the mixing cell model: $ud_p/D_a = \text{Pe} = 2$ (see Figure 6, top). But, strangely enough, measurements with liquid flow down to Reynolds numbers of about 10^{-2} also agree (Figure 6, bottom). At such low Reynolds numbers, however, there is no more eddy diffusivity, and the concept that the voids between the pellets act as mixing cells has no meaning in this region. Hence a different concept must be applied here.

Taylor Dispersion

Radial mixing and axial dispersion are not as independent of one another as it may seem from their different characteristics that were noted in the foregoing section. In fact, the two processes are interconnected by the mechanism of the Taylor dispersion. This mechanism also leads to an understanding of the considerable differences in axial dispersion between gas and liquid flow that are represented in Figure 6.

For a short introduction to the Taylor dispersion concept, let us imagine viscous flow in an open cylindrical tube, and let us consider the RTD of the flowing volume elements. This distribution function is represented by the solid curve in Figure 2. It starts with the highest probability at the smallest residence time—that which belongs to the fast flow near the tube axis and which is equal to half the mean value τ—and it decreases down to the long residence times of the slowly flowing volume elements near the wall. The broken line in Figure 2 represents the signal of a tracer pulse that was injected upstream uniformly over the cross-section. The difference in RTD's is brought about by radial mixing processes that act on the axial tracer profile but have no effect, of course, on the flow pattern.

The interference of radial mixing becomes clear from Figures 3 and 4. Figure 3 depicts the axial dispersion of the tracer pulse as it would develop under the effect of only the local differences in flow rate according to the velocity profile. Such a spread of the tracer probe tends to build up boundaries with steep concentration gradients in the radial direction (which is marked by the arrows in Figure 3). In fact, however,

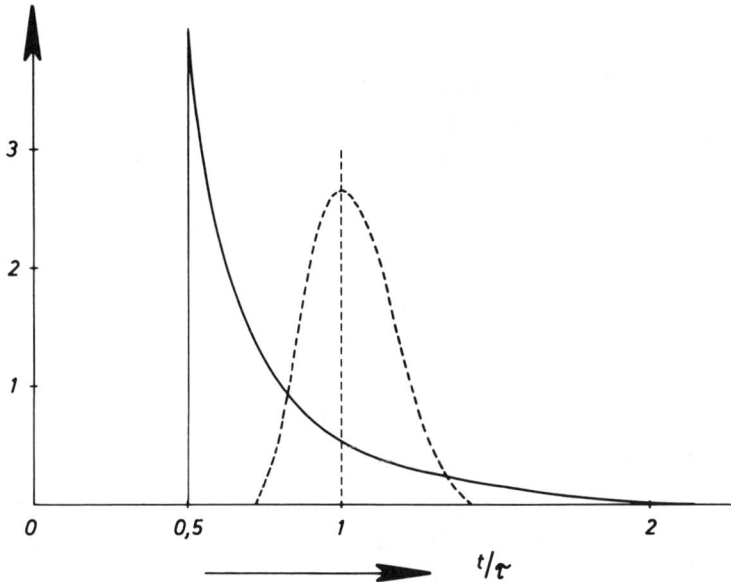

Figure 2. Residence time distribution of laminar flow in open tube
$$\frac{d(m/m_0)}{d(t/\tau)} = \frac{1}{2(t/\tau)^3}$$
..., tracer pulse with radial mixing, schematic

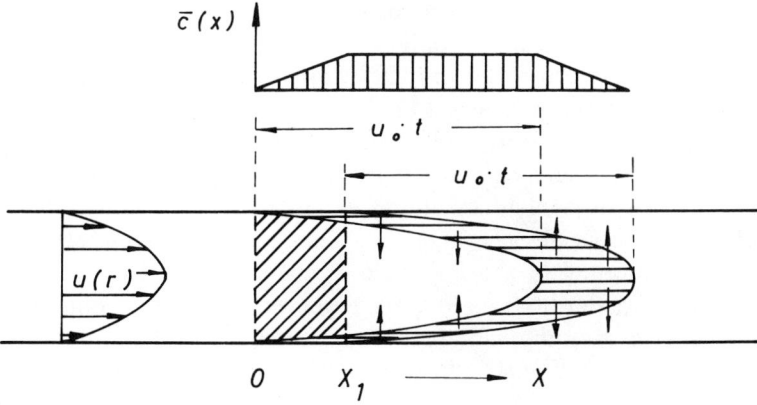

Figure 3. Axial dispersion of tracer pulse without radial diffusion

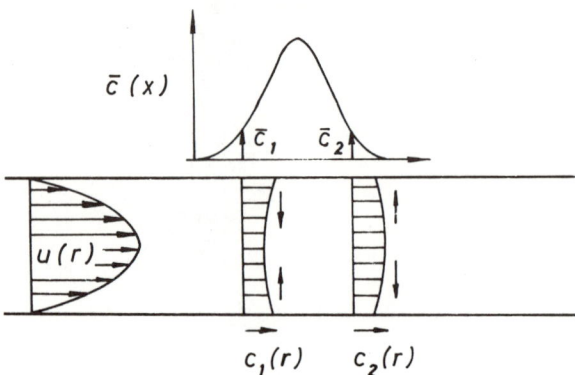

Figure 4. Narrowing of axial dispersion by radial diffusion: Taylor dispersion

radial diffusion along these gradients transfers tracer material from the top of the pulse to regions of slower flow (*see* Figure 4) and from the wake of the pulse to regions of faster flow. In this manner, the extent of axial dispersion is diminished by radial mixing. The result is the development of a bell-shaped concentration profile (Figure 4) that is similar to a peak in chromatography. The spread of the profile or the variance σ^2 developed along a flow path of length L determines the Taylor dispersion coefficient by

$$\sigma^2 = 2D_T \cdot L/u \qquad (8)$$

D_T in turn is inversely proportional to the coefficient of radial mixing (*see* Equation 9). This concept, which was developed by Taylor (*15*) for flow in open tubes, can be applied readily to axial dispersion in packed beds. This was demonstrated by Edwards and Richardson (*16*) and by Bischoff (*17*); [*see also* Wicke (*12, 13*)].

The Taylor dispersion coefficient can be taken proportional to the mean flow rate, u, and to a mean travelling distance in the main flow direction, Λ, along which a volume element loses its individuality by lateral mixing processes: $D_T \sim u \cdot \Lambda$. The so-called mixing length Λ is equal to the product of the flow rate u and the mean lifetime τ of the volume element; the lifetime in turn is determined by the square of the pellet diameter over the radial mixing coefficient: $\Lambda = u \cdot \tau = u \cdot d_p^2 / M_r$. In this way, the characteristic expression

$$D_T \sim \frac{u^2 \cdot d_p^2}{M_r} \qquad (9)$$

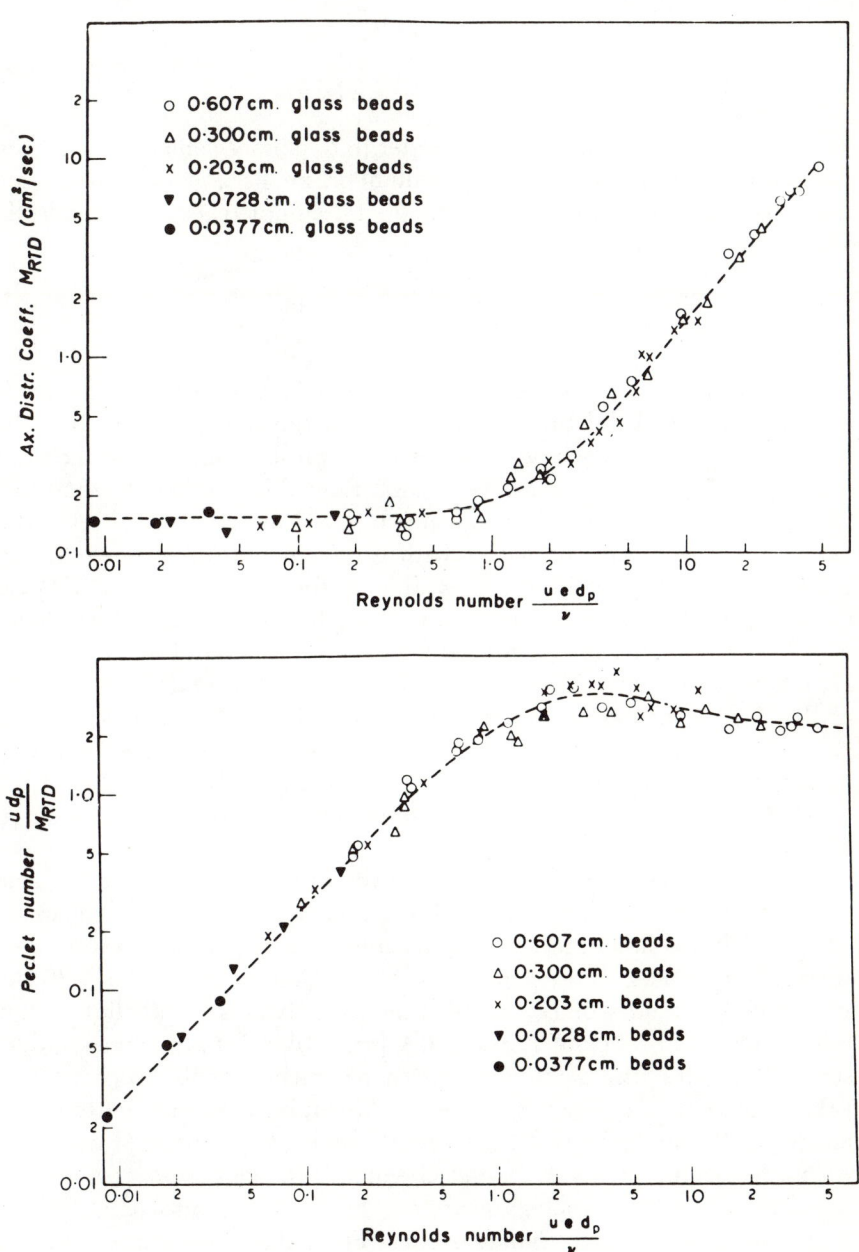

Figure 5. *Axial distribution in air flow through packed bed of glass beads* (16)

for the Taylor dispersion coefficient is obtained. Introduction of M_r for packed beds (Equation 6) gives

$$D_T = 0.5 \cdot \frac{ud_p}{1 + 10\chi D_m/ud_p} \qquad (10)$$

where the factor 0.5 was chosen in order to obtain agreement of Equation 10 with the limiting value of Peclet number, Pe = 2, at Re ≳ 10. When the molecular diffusion term is added, the complete expression for the axial distribution coefficient becomes

$$M_{RTD} = \chi D_m + 0.5 \cdot \frac{ud_p}{1 + 10\chi D_m/ud_p} \qquad (11)$$

This differs from the simpler Equation 7, that was based on the mixing cell model, by the additive term in the denominator. Just this term, however, is necessary for a correct representation of the measurements in the transition region from high Reynolds numbers to the range of molecular diffusion, as was shown by Edwards and Richardson (16). By careful measurements with argon pulses in air flow through packed beds of glass beads, they demonstrated that the axial Peclet number runs through a flat maximum in the transition region (see Figure 5). The maximum can be reproduced with Equation 11 but not with Equation 7; this thereby confirms the applicability and the usefulness of the Taylor dispersion concept for packed beds. With liquid flow in packed beds of spheres, the corresponding maximum has been observed earlier in profound tracer studies by Hiby (11) in the range $10^{-3} <$ Re $< 10^{-2}$; he represented his measurements by a formula similar to Equation 11.

Figure 6 gives a more extended survey of experimental data on axial Peclet numbers collected from the literature by Edwards and Richardson (16). With gas flow, the curve splits into several parallel lines at small Reynolds numbers. This represents the region of molecular diffusion with different values of D_m in different gas mixtures. With liquid flow, this region occurs below Re $\approx 2 \cdot 10^{-3}$ [see Hiby (11)]; down to these small flow rates, the axial distribution of tracer signals keeps its high value that is characterized by Pe ≈ 2. An explanation may be found by means of the capillary model for flow through packed beds (Figure 1c). At the branching points A, B, and C, volume elements from regions near the axis pass over to regions near the wall; at the points of confluence D, E, and F, the reverse change occurs. From these random retardations and accelerations of volume elements arises the axial distribution of a tracer pulse that is injected into the flow. It must be postulated, however, that the tracer substance will not pass over laterally from one volume

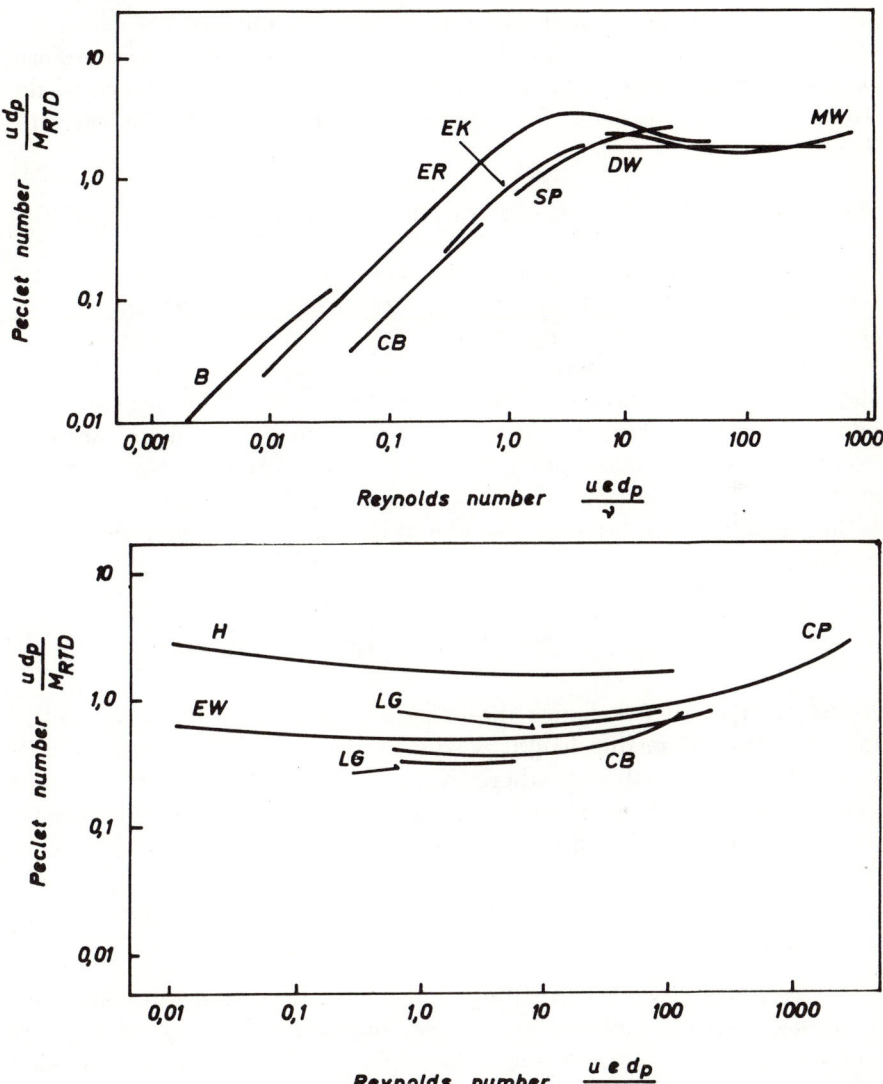

Figure 6. Peclet numbers for axial distribution in packed beds (16) with gas flow (top) and liquid flow (bottom)

Key for gas flow: B, Blackwell et al. (18); CB, Carberry and Bretton (19); DW, De-Maria and White (22); EK, Evans and Kenney (20); ER, Edwards and Richardson (16); MH, McHenry and Wilhelm (23); and SP, Sinclair and Potter (21). Key for liquid flow: CB, Carberry and Bretton (19); CP, Cairns and Prausnitz (26); EW, Ebach and White (24); H, Hiby (11); and LG, Liles and Geankoplis (25).

element to the other; thus radial mixing must be restricted. This is true for liquids with their small diffusion coefficients but not for gases; the high axial distribution at low Reynolds numbers is observed, therefore, only with liquid flow. This indicates once more the importance of the Taylor dispersion concept for heterogeneous flow systems [Wicke (12, 13)].

Taylor Dispersion with Mass Transfer between Phases: Fluidized Beds

If mass transfer between phases occurs in heterogeneous flow systems, the relaxation of this lateral transfer process gives rise to an additional type of axial dispersion. This effect is of particular importance in gas chromatography where mass transfer between the mobile and the stationary phase is involved [see e.g. Giddings (4)]. The principles are as follows. If for the mass transfer the diffusion in one phase, say in a liquid film, is rate-determining, then the mixing length Λ can be taken as $\Lambda = u \cdot \tau_{\text{diff}} = ul^2/D_{\text{eff}}$ where l is a characteristic distance of the diffusion problem, for instance the film thickness. The Taylor dispersion coefficient is then:

$$D_T \sim u \cdot \Lambda \sim \frac{u^2 l^2}{D_{\text{eff}}} \qquad (12)$$

If instead, the mass exchange through the phase boundary is rate-determining, then the mixing length is equal to the height of a transfer unit $\Lambda = \text{HTU} = u/(a \cdot K_{\text{exch}})$ where K_{exch} is the exchange rate constant (mass transfer coefficient) and a is the interphase area per unit volume of the bed. The Taylor dispersion coefficient then becomes:

$$D_T \sim u \cdot \Lambda \sim \frac{u^2}{a \cdot K_{\text{exch}}} \qquad (13)$$

The possibility of correlating interphase mass transfer with axial dispersion has been applied in chemical engineering in only a few cases so far. The case of interest here is the application to the mixing processes in fluidized beds that was developed by Levenspiel and co-workers (27). The principal relations for radial and axial dispersion of the solid particles and of the gas phase according to this development are summarized in Table II. The formulas for radial dispersion are based on the random walk model with the mean diameter of the bubbles, d_b, as the characteristic walking distance. The rate-determining processes were assumed to be the lateral mixing of the solid particles in the wakes behind the bubbles and the exchange of the gas between the bubbles and the emulsion phase. The formulas for the axial dispersion are based on the Taylor

Table II. Dispersion of Solids and Gas in Fluidized Beds (27)

Type	Concept	Phase	Equation
Radial	random walk	solid	$D_r^s = \dfrac{3}{16} \cdot \dfrac{\delta}{1-\delta} \cdot u_f d_b$ (14a)
		gas	$D_r^g = 0.2 \cdot \dfrac{d_b^2}{\delta/K^g_{exch}}$ (14b)
Axial	Taylor dispersion	solid	$D_a^s = \dfrac{\delta}{1-\delta} \alpha^2 (1-\epsilon_{mf}) \cdot \dfrac{u_b^2}{K^s_{exch}}$ (15a)
		gas	$D_a^g = f(u_o, u_f, \ldots) \dfrac{u_o u_b}{K^g_{exch}}$ (15b)

d_b = effective bubble diameter
δ = fraction of fluidized bed consisting of bubbles
α = ratio of wake volume to bubble volume
u_b = rising velocity of a bubble
u_o = gas velocity in empty tube
u_f = gas velocity under minimum fluidizing conditions
ϵ_{mf} = void fraction under minimum fluidizing conditions
K^g_{exch} = exchange coefficient of gas between bubbles and emulsion phase
K^s_{exch} = exchange coefficient of solids between the cloud-wake region and the emulsion phase

principle with interphase mass transfer; the mass exchange here occurs between the bubble phase or the wakes and the emulsion phase.

These formulas represent remarkable progress toward a systematic understanding of the mixing processes in bubbling fluidized beds; quantitatively, however, they yield only the orders of magnitude. It must be assumed that one reason for this is that a fluidized bed usually extends over just a few walking distances or mixing lengths, and therefore it falls just in the gap where the model of mixing cells is no longer valid and the concept of pseudo-homogeneous dispersion is not yet applicable. The instructive investigation by Nguyen and Potter (28) of the backmixing in fluidized beds of gas components that are adsorbed by the solid particles was done in this particular range.

A second reason for the quantitative failure of the formulas in Table II is the poor uniformity in bubble size, especially the growth of the bubbles during their rise through the bed. An interesting possibility for limiting bubble size was presented by Kato and co-workers (29); they are running a fluidized bed in the voids of a packed bed of coarse pellets, and they expect a number of advantages from this technique.

Generally, fluidized bed modelling suffers from a lack of reliable experimental data, as Kunii and Levenspiel (27) noted repeatedly. Kojima and co-workers (30) are to be welcomed, therefore, with their new electrochemical method for measuring the local velocities and directions of the dense phase, even though it is not in fluidized beds but in the closely related case of bubble columns.

Taylor Dispersion Concept for Interphase Heat Transfer

Besides the applications of the Taylor dispersion concept that were mentioned so far, additional applications will surely be developed in the future. Vortmeyer and Schaefer (*31*) recently called attention to the possibility that the relaxation of interphase heat transfer in heterogeneous gas–solid flow systems may be accounted for in a pseudo-homogeneous model by introducing an additional axial dispersivity of heat. Remembering that the HTU for heat transfer is: $\Lambda_h = (u \cdot \rho \cdot c_p)/(a \cdot h)$, the Taylor dispersion coefficient for interphase heat transfer can be taken in analogy with Equation 13 for mass transfer as

$$D_{T,h} \sim u \cdot \Lambda_h \sim \frac{u^2 \cdot \rho \cdot c_p}{a \cdot h} \tag{16}$$

Accordingly, the term: $\Delta\lambda = D_{T,h} \cdot \rho \cdot c_p$ must also be introduced into the expression for the effective axial heat conductivity in the gas–solid flow system. Vortmeyer and Schaefer followed a different line of approach and arrived at the corresponding formula:

$$\lambda_a = \lambda_o + \frac{(u \cdot \rho \cdot c_p)^2}{a \cdot h} \tag{17}$$

where λ_o is the effective heat conductivity in the packed bed with the gas phase at rest. In order to test the applicability of the method, the authors calculated the unsteady state problem of heating up a packed bed of glass beads by means of a hot air flow, and they compared the results with the exact two-phase calculations. A comparison of the temperature profiles calculated for warming-up times of 200 and 600 sec is presented in Figure 7. The profile calculated with the one-phase model coincides in good approximation with the profile of solids temperature from the two-phase calculations. The method was extended recently to computing the steep temperature profiles through moving reaction zones in packed beds; Vortmeyer and co-workers reported this application of the pseudo-homogeneous model (*32*).

Catalyst Screens and Shallow Beds

The investigation of chemical reaction and mass transfer at screen catalysts, performed with hydrogen peroxide decomposition by Shah and

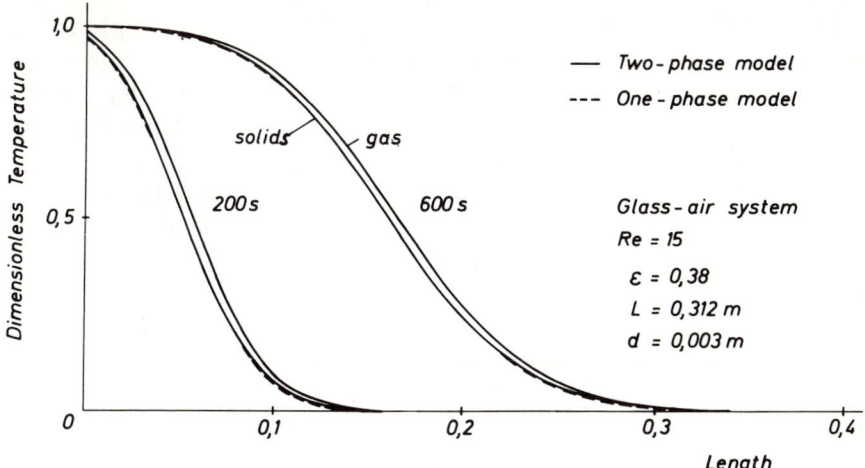

Figure 7. Heating-up of a packed bed by hot air flow. Calculation with one-phase model based on the Taylor dispersion concept for interphase heat transfer (31).

Roberts (33), seems to be outside the line of this review on pseudo-homogeneous modelling. Actually, however, it represents a limiting case of a heterogeneous flow system that is comparable to a shallow packed bed with one or a few layers of pellets. The objective of the investigation was to extend the earlier studies of Satterfield and Cortez (34), who worked in the range of $0.5 < \text{Re} < 10$, to higher Reynolds numbers (up to $\text{Re} \approx 100$). The velocities of the gas flow—nitrogen with 0.3–2% H_2O_2 vapor—are accordingly high, 1–10 m/sec, and for this reason the authors neglect the effect of longitudinal diffusion. However, if noticeable conversion occurs when the gas flow passes a screen—the authors used single screens and stacks of 3, 4, and 5 screens—there must be appreciable diffusion transversal to flow direction in order to achieve the mass transfer to the surface of the wires. The penetration depth of the transversal diffusion into the gas stream that passes a mesh of the screen can be assumed to be of the order of the wire radius.

A similar order is to be expected for the longitudinal backdiffusion. Actually, the authors reported D_m/u values of ≤ 0.004 cm for the depth of backmixing under reaction conditions whereas the wire diameters of the used screens are 0.008–0.035 cm. The effect of longitudinal diffusion on the mean concentration gradient in the flow direction between the wires may be estimated therefrom as is shown in Figure 8 by the dotted line. The driving force is diminished by Δc_i at the inlet and by Δc_e at the exit of the mesh.

The authors, on the other hand, assumed plug flow; they evaluated the measurements by means of the logarithmic mean driving force, and correlated the data in analogy with heat transfer to the single infinite cylinder under cross-flow conditions. With regard to Figure 8, however, a treatment in analogy with heat transfer to a tube wall under entrance flow conditions seems more realistic; in fact, this type of modeling was applied long ago by Wagner (35).

The analogy with entrance flow to a tube is advantageous also because it is equally applicable to screen catalysts and to shallow packed

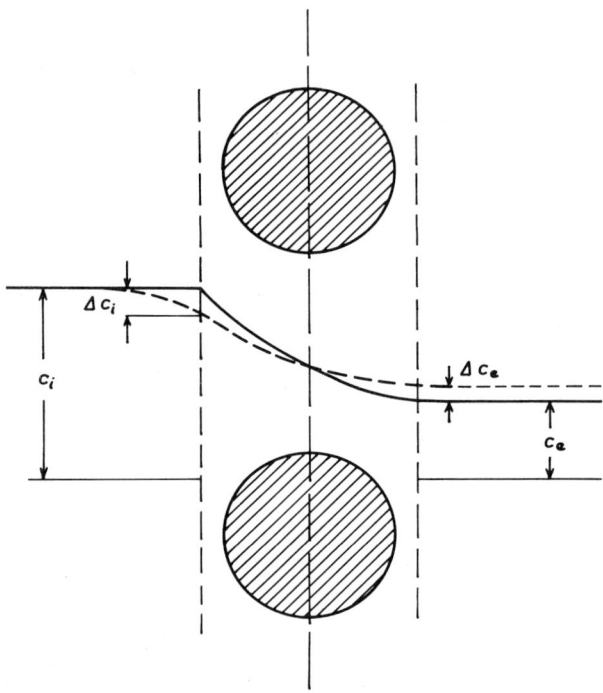

Figure 8. Effect of longitudinal diffusion on the mean concentration profile between the wires (gas flow is from left to right)

beds. Along this line, the transition from a single layer of pellets to several layers and finally to a packed bed could be investigated both theoretically and by experiment. Strangely enough, this transition has been given little attention so far; if one asks why, the answer is usually: because there was no need for this transition range in chemical reaction engineering. The situation has changed, however, in recent years; shallow packed beds of pellets or monoliths became the prototypes of catalytic mufflers for removal of CO and NO from exhaust gases [Hegedus (36), Szepe (37)]. This challenge will surely increase the efforts to fill the gap in our knowledge about the above-mentioned transition range.

The Problem of Boundary Conditions for the Fixed Bed Tubular Reactor

Imagine a catalytic fixed bed reactor with a single reaction at steady state, the catalyst section being sufficiently long to permit application of the pseudo-homogeneous model for axial dispersion of the fluid flow between the pellets. The dispersion diminishes the progress of chemical conversion along the flow path. In case of linear reaction kinetics—first-order with reference to a key component i—this effect can be described by a diffusionlike term in the mass balance equation of this component. The balance per unit volume of fluid between the pellets is

$$\frac{d}{dx}\left(M_{\text{RTD}} \frac{dc_i}{dx}\right) - \frac{d}{dx}(uc_i) + v_i \frac{1-\epsilon}{\epsilon} \cdot v_{\text{eff}} = 0 \tag{18}$$

with M_{RTD} as in Equation 11. This balance corresponds to the axial-dispersed plug-flow model of Levenspiel and Bischoff (38, 39, 40) [see also Wilhelm (10)]; the neglect of radial variables means that the reaction runs under isothermal or under adiabatic conditions, so no radial gradients of temperature and of concentration have to be considered. The boundary conditions are usually taken as those proposed by Danckwerts (41) and generalized by Wehner and Wilhelm (42):

$$c_{oi} = c_i(0^+) - \frac{D}{u}\left(\frac{dc_i}{dx}\right)_{0^+} \tag{19a}$$

$$\left(\frac{dc_i}{dx}\right)_{L^-} = 0 \tag{19b}$$

where c_{oi} is the concentration of component i in the feed (at $x \to -\infty$), and where 0^+ refers to a position immediately downstream of the entrance cross-section of the catalyst bed at $x = 0$ and L^- to a position immediately upstream of the exit at $x = L$. The condition in Equation 19a accounts for the effect of longitudinal diffusion that reduces the concentration in the entrance by the amount:

$$\Delta c = c_{oi} - \lim_{x \to 0} c_i(0^+) \tag{20}$$

below the feed concentration c_{oi}. The condition in Equation 19b results from the argument that no discontinuity in the gradient should occur at the exit of the catalyst bed, i.e.

$$\left(\frac{dc_i}{dx}\right)_{L^-} = \left(\frac{dc_i}{dx}\right)_{L^+}$$

and $(dc_i/dx)_{L^+} = 0$ because chemical conversion vanishes downstream of the exit.

There has been much discussion about these boundary conditions, especially about Equation 19a. Pearson (43) discussed the apparent jump in concentration at the entrance, Equation 20. van der Laan (44) and van Cauwenberghe (45) investigated the applicability of the boundary conditions to unsteady state processes. Bischoff (46) demonstrated their validity in the case of steady state for any type of reaction kinetics other than first-order. Two aspects, however, have apparently not been considered up to now: the nature of the diffusion coefficient D in Equation 19a and the entrance effects on RTD when a flow enters a packed bed from an empty tube. These aspects are discussed briefly in the following sections.

Case I. The reactor tube upstream of the catalyst section may be filled with inactive pellets of the same size and shape as the catalyst pellets (*e.g.* inactive support material). At the entrance to the catalyst bed, the characteristic pattern of packed bed flow is then fully established. Hence the dispersion processes are also fully developed, and the value of M_{RTD} as calculated by Equation 11 can be used in Equation 18 without suspicion. The same is not true, however, for the coefficient D in Equation 19a although this coefficient is usually identified with M_{RTD}. Actually this coefficient represents a diffusion with reference to a plane fixed in space—the entrance cross-section of the catalyst bed at $x = 0$— *i.e.* a real backmixing process, but no dispersion effects. Therefore only the backmixing term of Equation 11 must be used in Equation 19a: $D = \chi \cdot D_m$. The depth of penetration of this longitudinal backdiffusion, Δx, can be estimated as:

$$\frac{\Delta x}{d_p} = \frac{\chi D_m}{u d_p} < \tfrac{1}{2} \qquad \text{for Re} > 1 \qquad (21a)$$

where the numerical values were obtained from Figure 6. Hence the penetration depth is only a fraction of pellet diameter even at small Reynolds numbers, and it diminishes rapidly with increasing flow rate. Likewise the concentration difference at the entrance, Δc in Equation 20, is only a fraction of the decrease in concentration along one pellet diameter

$$\frac{\Delta c}{\left(\dfrac{dc_i}{dx}\right)_{0+} \cdot d_p} \approx \frac{\Delta x}{d_p} \qquad (21b)$$

that diminishes in the same way with increasing flow rate. On the other hand, the pellet diameter is the smallest distance that can be worked with in the pseudo-homogeneous model of a packed bed; the penetration depth in Equations 21a and 21b therefore has no meaning in this model. Accordingly, for the case discussed here, the boundary condition (Equation 19a) can be simplified to

$$c_{oi} = c_i(0^-) = c_i(0^+) \qquad (22)$$

as was proposed earlier by Hulburt (*47*). This conclusion means that the mass balance (Equation 18), wherein the axial dispersion is accounted for, must be solved in accordance with the boundary condition for piston flow (Equation 22) at the entrance to the catalyst bed. For mathematical reasons, it may be suitable to use the complete condition (Equation 19a); then only the molecular diffusion must be considered, i.e. $D = \chi \cdot D_m$.

Case II. The reactor tube upstream of the catalyst section may be empty so the flow entering this section at $x = 0$ enters the packed bed at the same time. Passage through 3–5 particle layers is needed in order to develop the twisted pattern of interparticle flow. This entrance zone contributes more to the broadening of the RTD than other zones downstream in the bed; the reason is the acceleration from mean flow rate u_o in the empty tube to $u = u_o/\epsilon \approx 2.5 \cdot u_o$ between the particles, which results in appreciable differences in the velocity of adjacent volume elements (before the transversal components of flow with their averaging effects have developed fully). Actually, the predominant part of this entrance dispersion process occurs upstream of and in the first particle layer. Part of the streamlines of the empty tube flow are drawn immediately into the interstices between the pellets and thereby accelerated; others are dammed up in front of single pellets and thereby retarded. Within this entrance zone, higher values for M_{RTD} should be used in Equation 18. Alternately, if the constant value from Equation 11 is taken, the length of the catalyst section must be extended in the calculations to an effective value $L_{eff} = L + \Delta L$ where ΔL accounts for the additional broadening of the RTD by the entrance effects. The value of ΔL is expected to be on the order of a few particle diameters (and proportional to d_p). As the entrance boundary condition in this case, Equation 22 should be applied at the position $x = -\Delta L$. This boundary condition, however, cannot be used at present because the dependence of ΔL on flow rate and on reaction kinetics is not known. Elucidation of the proper relationships is a problem for future research.

Nomenclature

a	= interphase area per unit bed volume
c	= concentration in fluid phase

c_p	= specific heat of fluid
D_a, D_r	= axial and radial dispersion coefficients
D_{eff}	= effective diffusion coefficient
D_m	= diffusion coefficient in free gas phase
d_p	= particle diameter
D_T	= Taylor dispersion coefficient
h	= heat transfer coefficient
K_{exch}	= exchange rate coefficient
l	= characteristic length
L	= length of a packed bed section
M_r	= radial mixing coefficient
M_{RTD}	= axial distribution coefficient
n_a	= amount adsorbed (or absorbed) at the pore walls per unit volume of porous body
Pe	= ud_p/M_{RTD} = Péclet number
r	= pore radius
Re	= ud_p/ν = Reynolds number
RTD	= residence time distribution
t	= time
u	= mean axial flow velocity in packed bed
u_o	= mean axial flow velocity in empty tube
v_{eff}	= reaction rate per unit volume of porous support
x	= space coordinate in axial direction

Greek

ϵ	= void fraction in packed bed
ϵ_p	= porosity
λ	= heat conductivity
Λ	= mixing length
ν_i	= stoichiometric number of reaction component i
ρ	= mass density of fluid
σ^2	= variance of pulse profile
τ	= mean residence time, mean lifetime
χ	= labyrinth factor ($1/\chi$ = tortuosity)
ψ	= permeability of porous body

Literature Cited

1. Wakao, N., ADV. CHEM. SER. (1974) **133**, 281–289.
2. Abed, R., Rinker, R. G., *J. Catal.* (1973) **31**, 119.
3. Livbjerg, H., Sorensen, B., Villadsen, J., ADV. CHEM. SER. (1974) **133**, 242–258.
4. Giddings, J. C., "Dynamics of Chromatography," Part I, Marcel Dekker, New York, 1965.
5. Hugo, P., *Chem. Eng. Sci.* (1965) **20**, 187.
6. *Ibid.* (1965) **20**, 385.
7. *Ibid.* (1965) **20**, 975.
8. Hesse, D., *Ber. Bunsenges. Phys. Chem.* (1974) **78**, 744.
9. *Ibid.* (1974) **78**, 753.
10. Wilhelm, R. H., *Pure Appl. Chem.* (1962) **5**, 403.
11. Hiby, J. W., "Interactions between Fluids and Particles," *Proc. Symp. Inst. Chem. Eng.*, London, 1962, p. 312.

12. Wicke, E., *Ber. Bunsenges. Phys. Chem.* (1973) **77**, 160.
13. *Ibid.* (1965) **69**, 761.
14. Aris, R., Amundson, R. N., *AIChE J.* (1957) **3**, 281.
15. Taylor, G., *Proc. Roy. Soc. London Ser. A* (1953) **219**, 186.
16. Edwards, M. F., Richardson, J. F., *Chem. Eng. Sci.* (1968) **23**, 109.
17. Bischoff, K. B., *Chem. Eng. Sci.* (1969) **24**, 607.
18. Blackwell, R. J., Rayne, J. R., Terry, W. M., *J. Pet. Technol.* (1959) **11**, 1.
19. Carberry, J. H., Bretton, R. H., *AIChE J.* (1958) **4**, 367.
20. Evans, E. V., Kenney, M. A., *Trans. Inst. Chem. Eng.* (1966) **44**, T189.
21. Sinclair, R. J., Potter, O. E., *Trans. Inst. Chem. Eng.* (1965) **43**, 3.
22. DeMaria, F., White, R. R., *AIChE J.* (1960) **6**, 473.
23. McHenry, K. W., Wilhelm, R. H., *AIChE J.* (1957) **3**, 83.
24. Ebach, E. E., White, R. R., *Chem. Eng. Sci.* (1958) **4**, 161.
25. Liles, A. W., Geankoplis, C. J., *AIChE J.* (1960) **6**, 591.
26. Cairns, E. J., Prausnitz, J. M., *Chem. Eng. Sci.* (1960) **12**, 20.
27. Kunii, D., Levenspiel, O., "Fluidization Engineering," John Wiley, New York, 1969.
28. Nguyen, H. V., Potter, O. E., ADV. CHEM. SER. (1974) **133**, 290–300.
29. Kato, K., Arai, H., Ito, U., ADV. CHEM. SER. (1974) **133**, 271–280.
30. Kojima, E., Akehata, T., Shirai, T., ADV. CHEM. SER. (1974) **133**, 231–241.
31. Vortmeyer, D., Schaefer, R. J., *Chem. Eng. Sci.* (1974) **29**, 485.
32. Vortmeyer, D., Dietrich, K. J., Ring, K. O., ADV. CHEM. SER. (1974) **133**, 588–599.
33. Shah, M. A., Roberts, D., ADV. CHEM. SER. (1974) **133**, 259–270.
34. Satterfield, C. N., Cortez, D. H., *Ind. Eng. Chem. Fundam.* (1970) **9**, 613.
35. Wagner, C., *Chem. Technik* (1945) **18**, 1.
36. Hegedus, L. L., "Effects of Channel Geometry on the Performance of Catalytic Monoliths," *Am. Chem. Soc., Div. Pet. Chem., Prepr.* (1973) **18**, 487.
37. Szépe, S., "The Optimal Geometric Structure of Catalytic Monolith Reactors," *Prepr. GVC/AIChE Joint Meeting, Munich, 1974*, Vol. I.
38. Bischoff, K. B., Levenspiel, O., *Chem. Eng. Sci.* (1962) **17**, 245.
39. *Ibid.* (1962) **17**, 257.
40. Levenspiel, O., Bischoff, K. B., "Advances in Chemical Engineering," Vol. 4, p. 95, Academic, New York, London, 1963.
41. Danckwerts, P. V., *Chem. Eng. Sci.* (1953) **2**, 1.
42. Wehner, J. F., Wilhelm, R. H., *Chem. Eng. Sci.* (1956) **6**, 89.
43. Pearson, J. R. A., *Chem. Eng. Sci.* (1959) **10**, 281.
44. van der Laan, E. T., *Chem. Eng. Sci.* (1958) **7**, 187.
45. van Cauwenberghe, A. R., *Chem. Eng. Sci.* (1966) **21**, 203.
46. Bischoff, K. B., *Chem. Eng. Sci.* (1961) **16**, 131.
47. Hulburt, H. M., *Ind. Eng. Chem.* (1944) **36**, 1012.

RECEIVED December 4, 1974.

5

Industrial Process Models—State of the Art

VERN W. WEEKMAN, JR.

Mobil Research and Development Corp., Research Dept., Paulsboro, N.J. 08066

> *Kinetic process models have come of age and are being used increasingly in process development, design, and operation. The current state of the art in the building and use of such reactor process models is reviewed. Few, if any, complete process models have been published; however, various incomplete pieces have appeared recently. From these pieces, it is possible to assess current trends and problems. Particular attention is given to kinetically lumping complex reaction mixtures along with attendant problems and benefits. In addition, recent findings on particle heat and mass transfer effects, fluid bed comparisons, and two-phase flow-packed bed reactors are reviewed. Finally, remaining problem areas in the development of process models are addressed.*

Process models used in the design, operation, and optimization of industrial reactors are the *raison d'etre* of reaction engineering. They represent its final product and the vehicle by which the body of reaction engineering theory is applied. In this review, the state of the art of process models is judged primarily from recently published pieces of industrial models. The word pieces is appropriate since, to this reviewer's knowledge, few if any complete process models have been published. Typically, a model is published with the rate constants missing, or without mention of the mixing patterns that occur in the reactor, or with the chemistry not identified specifically. This is understandable in view of the usefulness of industrial process models to reactor design and operation. Complete models will be published eventually, but, since they were developed only during the last 5–10 years, apparently more time must pass before any complete, though obsolete, models will be published. By looking between the lines, however, we can make some judgments as to the current state of the art. It is also well to keep in mind that those pieces already published are probably from earlier versions of the currently used model and that they do not represent the latest model that is actually being used in the field.

Selected academic contributions, which in this reviewer's opinion bear directly on the current state of the art, were included in this review. With the emphasis on the present state, many valuable academic contributions which may have a large impact on future generations of models were not included. Most of the reviewed papers were published within the last three years, most since the last International Reaction Engineering Symposium in Amsterdam. Finally, at the end of the review, research areas which could greatly improve the current state of the art are discussed.

The Compleat Process Model

It is important to remember what the industrially useful process model should contain. The complete process model (a) accounts for the effects of the full range of process variables (*e.g.* pressure, flow rates, and temperature) on product yields and properties; (b) allows prediction of the effect of a wide range of charge stock composition on product yields and properties; (c) predicts the effects of catalyst aging and of changes in catalyst properties on activity and selectivity; (d) encompasses the effects of process variables on mixing and on hydrodynamic phenomena; and (e) has been verified by extensive pilot plant or commercial tests. Product properties are usually the most difficult to quantify in terms of basic chemical or physical phenomena. For example, the omnipotent octane number in the petroleum industry is difficult to characterize in a basic sense.

Ideally, the process variables should be linked to the yields and properties in terms of fundamental physiochemical phenomena. In practical models it is not always possible to describe each effect in its most fundamental form, and correlations involving adjustable parameters are usually required. Basic fundamental relationships for all variables may be expensive to ascertain, and Prater's optimum sloppiness principle (*1*) must be invoked. This principle is illustrated in Figure 1 where the fundamentalness of the model is plotted *vs.* its usefulness and cost. The fundamentalness parameter has been roughly quantified in terms of the number of phenomenological laws divided by the number of adjustable parameters. When this ratio is zero, we have a purely correlative model whereas, when it is infinite, we have a purely theoretical model which contains no adjustable coefficients. While the correlative model is cheap, you get what you pay for, and its extrapolative properties are poor. On

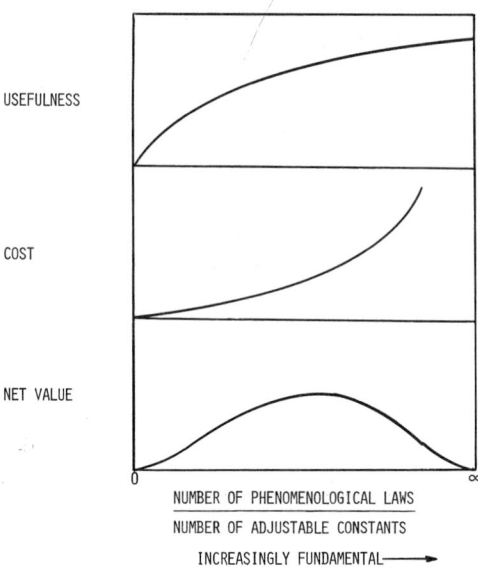

Figure 1. Principle of optimum sloppiness

the other hand, a purely theoretical model that explains all phenomena may give accurate extrapolations but at an exorbitant development cost. Thus, we see that the net value will probably have an optimum somewhat short of all phenomena being quantified. It is usually best to keep the model as simple as possible and to add only phenomena which contribute significantly to an understanding of the process variable behavior.

To return to our description of the complete process model, it is critical that it be able to predict the product distribution from the reaction over the full range of anticipated charge stock composition. Few industrial processes have single-component feedstocks, and, more often than not, a complex mixture must be reacted. It is very difficult to treat the reaction of each species, and by necessity we must lump species together kinetically. A substantial part of the review is devoted to the state of the art of kinetic lumping since it is so vital to most process models.

The next critical attribute of our industrial process model is the ability to predict the effects of catalyst aging and changes in catalyst properties on activity and selectivity. Once we know the rate constants for a given reaction scheme, we have the full power of the many tools of reaction engineering to aid us in design and operation. Unfortunately, we have few guidelines for predicting these rate constants from the properties of the catalyst. As a consequence, enormous sums of money are spent each year by industry for redetermining kinetic parameters after only small changes in catalyst properties that result from either aging or changes in composition.

Another key part of our process model is the ability to predict the effect of process variables on mixing and fluid dynamic phenomena. In this reviewer's experience, a significant portion of the difficulties encountered in applying industrial models lies in not fully understanding these phenomena. Thus changes in temperature or flow rate can, in turn, change mixing patterns so as to alter the reaction greatly. Finally, our model should be verified, at the very least in extensive pilot plant work or, more desirably, in large scale commercial tests. Only by such large scale testing can we be sure that the model successfully predicts the scale-up of all the key phenomena. Many times such large scale testing leads to the discovery of previously unaccounted for phenomena. Indeed, important discoveries of critical phenomena were made by the failure of process models in such tests. Prater (2) called this the strategy of failure; when applied alertly, it can be highly useful in sorting out the critical behavior of the process reactor.

Lumped Kinetics in Recent Process Models

One of the key problems in describing the kinetics of complex systems is how to lump the many components so that the resultant lumped kinetics describe the system behavior adequately. Some of the earliest theoretical work describing the nature of lumped systems was by Aris and Gavalas (3) and Aris (4). In a comprehensive treatment, Wei and Kuo (5, 6) described the errors involved in lumping monomolecular systems. As a matter of practicality, most industrial systems are strongly constrained to lumping those species which can be readily identified. There is also strong incentive to lump species in terms of those that are the final products of the process.

Catalytic Reforming. In the area of the catalytic reforming of petroleum naphthas, Smith (7) was one of the first to present lumped kinetics. His reaction scheme treated aromatics, naphthenes, and paraffins as single components. Each lumped species contained many compounds which were very likely to have different reaction rates. In spite of this, the kinetics were adequate to describe the overall behavior of the reformers. Recently Dorokhov *et al.* (8) also successfully described reformer behavior using a modification of this scheme (Figure 2). Unfortunately, this lumping is so coarse that it is sometimes difficult to describe the important properties of the system (*e.g.* octane). Here the distribution of compounds among the paraffins, naphthenes, and aromatics becomes important.

This shortcoming was rectified by Kmak (9) who described a lumped system for reforming that contained 22 lumped species. His basic

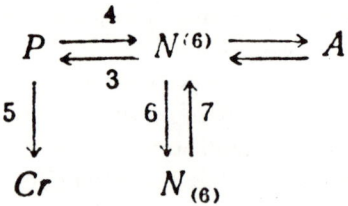

Khimiya i Tekhnologiya Topliv i Masel
Figure 2. Lumped kinetics of reformer behavior (8)

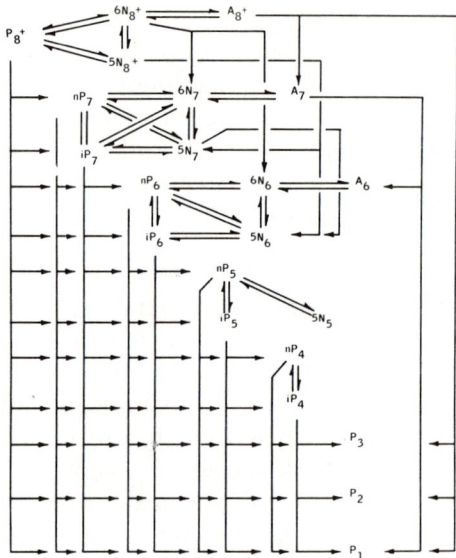

Figure 3. Kinetic model reaction paths
(9)

lumping scheme is depicted in Figure 3. While many of the lower molecular weight species are pure components, by the time one reaches the seven- and eight-carbon systems many different species are contained within each lump. Of course, such a system describes the overall chemistry and properties much more closely than the Smith model. Kmak was able to determine some of the rate constants independently although coupled fitting was required for a significant number of the constants. Such coupled fittings can distribute errors among the rate constants and makes hazardous extrapolation to regions not covered by the data. Figure 4 is Kmak's comparison of the predicted selectivity behavior with the pilot plant data. Increasing octane number reflects increased severity of operation and indicates that the model adequately describes the composition profiles for a number of individual components. Figure 5 compares the aromatics prediction with numerous pilot plant data. Kmak also compared the model with commercial data; again there was a very good fit.

Kmak's process model contains a heat balance term which allows prediction of temperature profiles in the bed. These calculations are compared with pilot plant data in Figure 6 which reveals that the model adequately describes the temperature profile through the reactors. Tests of both the heat and mass balanced equations are important in any thorough test of the model. Unfortunately, Kmak did not report his constants; however, one could reconstruct them with sufficient experimental data.

With this type of model, where a large number of rate constants are being fitted, it is extremely important that as many rate constants as possible be determined independently from the data. The number of rate constants which are fitted in a coupled fashion should be minimized in order to improve the predicting ability of the model, particularly in the region outside the range of the experimental data.

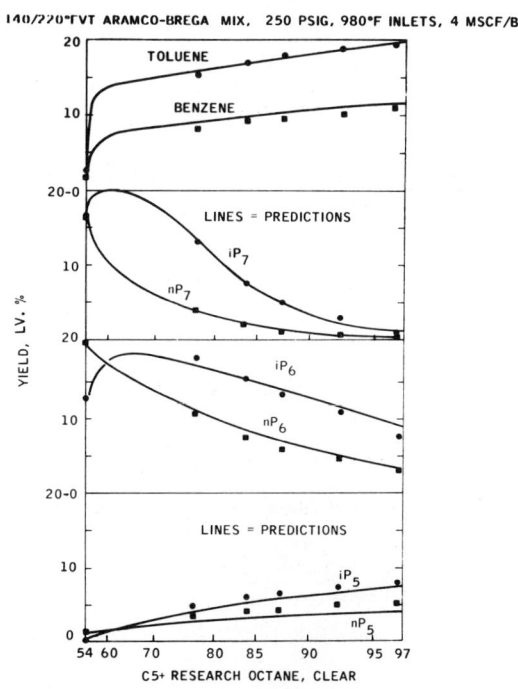

Figure 4. Pilot plant data and component profile predictions of selectivity behavior (9)

Hydrocracking Process Model. Hydrocracking of petroleum fractions is effected in order to reduce the molecular weight of a charge stock as well as to remove sulfur and other contaminants. The boiling range is usually considerably higher than with reforming—approximately 221°–482°C (430°–900°F). Because of the more complex molecular structures present, it is much more difficult to identify as many individual components as in reformers. As a result, one is forced to lump the individual constituents into much larger classes than in reforming.

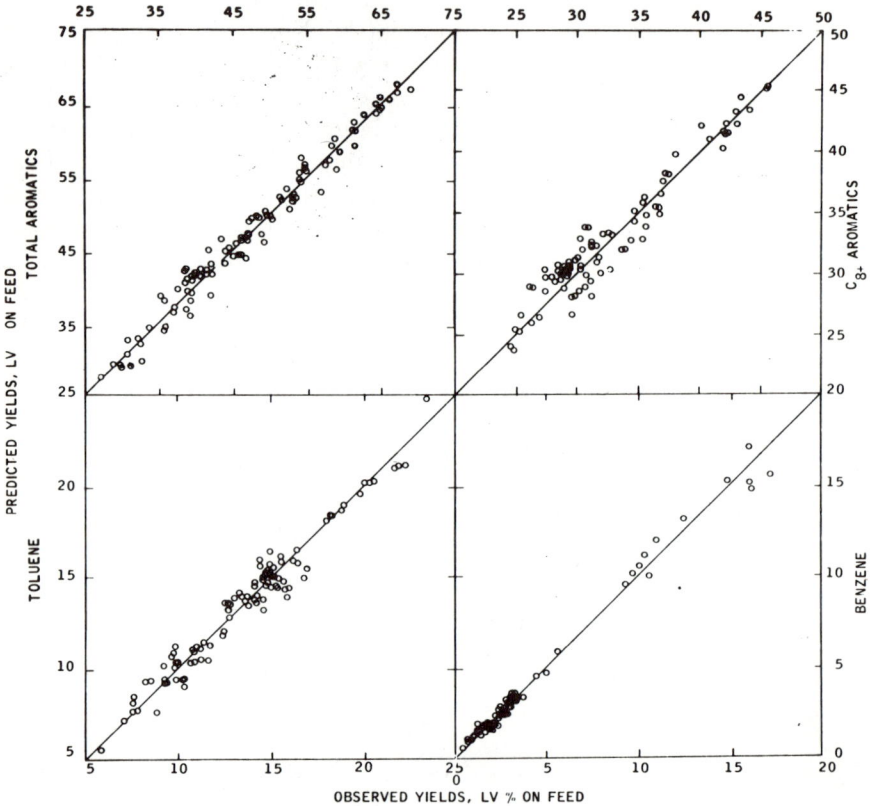

Figure 5. Pilot plant data and predictions of aromatic yields (9)

One approach by Stangeland (*10*) uses small boiling range fractions as the lumped components. In Figure 7 he plots the distribution of charge stock with boiling range for the discreet components chosen. Experimental data on the whole charge stock and the disappearance of the individual fractions were plotted in a first-order fashion (Figure 8). The rate constants obtained in this fashion were then correlated by the relationship

$$k(T) = k_o T + A(T^3 - T) \tag{1}$$

where the constants k_o and A are adjustable parameters which were determined from the experimental data. The relationship of A and k_o is depicted in Figure 9. Parameter A changes slightly with charge stock (Figure 10). Other similar parameters (B and C) used for the selectivity behavior, change at a greater rate with the paraffin concentration of the feed.

All the adjustable parameters will, of course, be strong functions of the particular catalyst used. Stangeland compared his predictions with

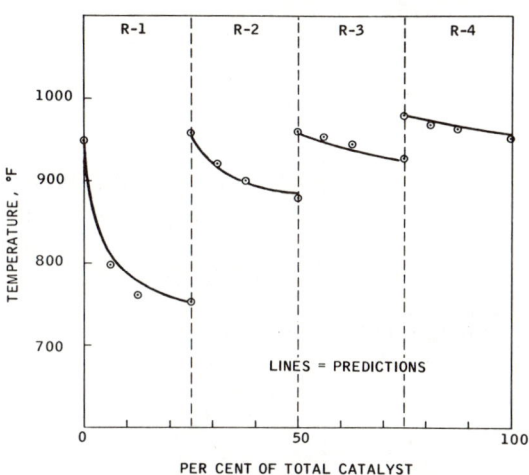

Figure 6. Pilot plant data and predictions of adiabatic reactor temperatures (9)

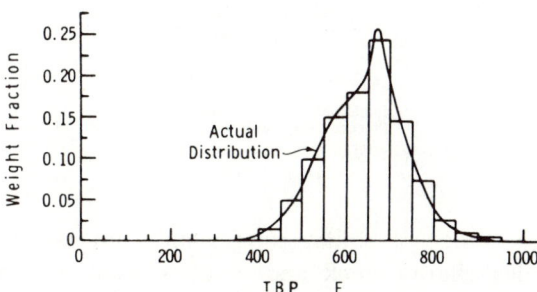

Industrial and Engineering Chemistry, Process Design and Development

Figure 7. Description of a feedstock as discrete components (10)

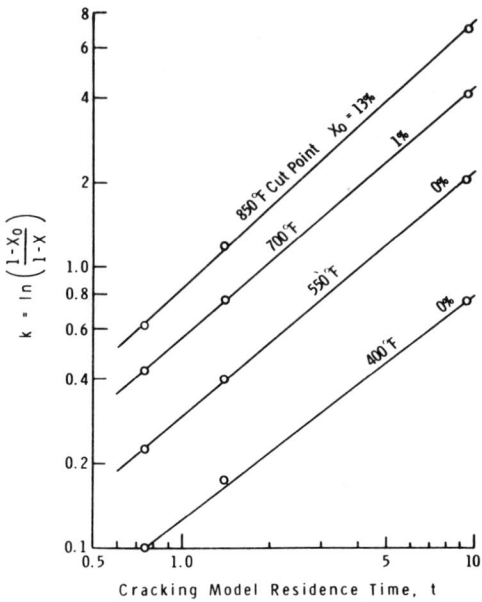

Figure 8. Correlation of first-order rate constant with model residence time (10)

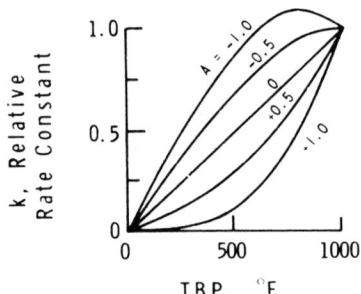

Figure 9. Cracking rate function (10)

data over a wide range of boiling ranges and conversions (Figure 11); for the particular charge stock and catalyst used, the model gave a good representation of the pilot plant data. Figure 12 is Stangeland's comparison of the model predictions with the jet fuel yields from a commercial two-stage hydrocracker; again the match between the commercial data and the predictions of the process model is very good.

Whereas the model is based on a kinetic framework, the correlations of the rate constants are not based on any fundamental phenomena. For this reason, considerable caution is required when this model is extrapolated into compositions of charge stock which were not included in the original experimental program.

Catalytic Cracking Models. The catalytic cracking of petroleum fractions uses a feedstock which is as complex as that used for hydro-

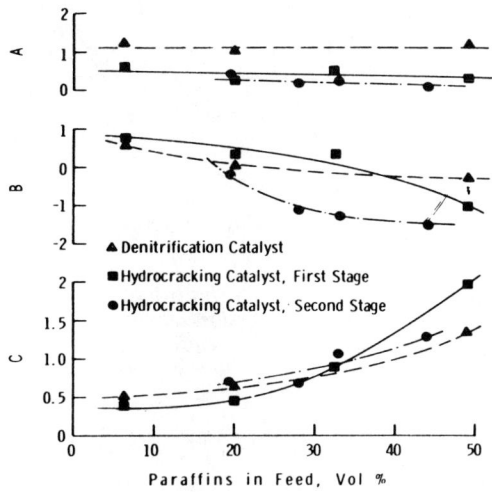

Industrial and Engineering Chemistry, Process Design and Development

Figure 10. The dependence of model parameters on feed and catalyst type (10)

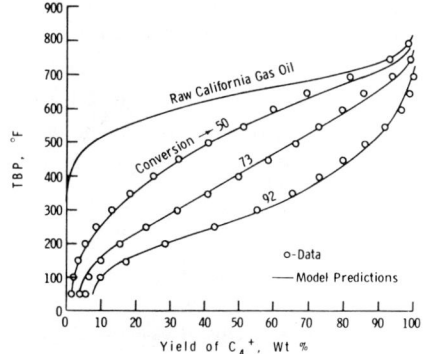

Industrial and Engineering Chemistry, Process Design and Development

Figure 11. Measured and predicted yields for through hydrocracking (10)

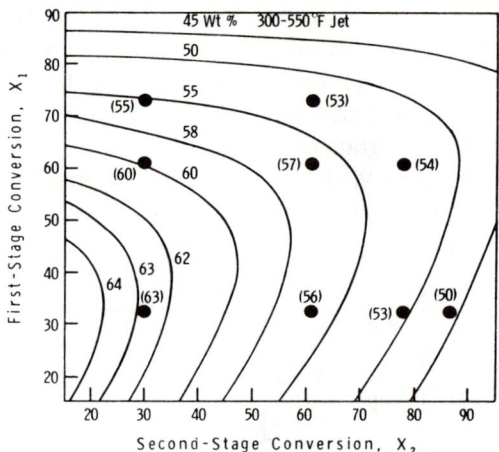

Industrial and Engineering Chemistry,
Process Design and Development

Figure 12. Predicted and measured jet yields from a two-stage hydrocracker showing the effect of conversion levels (10). Solid symbols indicate data; the jet yield is in parentheses; contour lines represent model predictions.

cracking. Weekman and co-workers (11, 12) described the kinetics of catalytic cracking in terms of the three-lump system diagrammed below.

$$\text{gasoline} \leftarrow \text{charge} \rightarrow \text{coke} + \text{light gases}$$

Here the charge stock is included in one lump with the major products constituting the other two lumps. Wojciechowski and co-workers (13, 14) also used the same lumping scheme. The two kinetic relationships describing the conversion of the charge stock are given in Equation 2 (Weekman, et al.) and in Equation 3 (Wojciechowski, et al.) for an isothermal fixed bed reactor.

$$X_A = \frac{1}{\dfrac{S}{K_o} t^n + 1} \qquad (2)$$

where K_o = reaction rate parameter, n = decay parameter (N in Equation 3), t = catalyst exposure time, and S = space velocity.

$$\frac{K_o}{S(1 + Gt)^{-N}} = \int_0^{X_A} \left(\frac{1 + \epsilon_A X_A}{1 - X_A}\right)^{1+W} dX_A \qquad (3)$$

The key difference in the two approaches is the two additional parameters in Wojciechowski's equation: G represents an additional decay parameter, the other a charge stock refractoriness factor (W). While Equation 3 may give a slightly better fit to experimental data, this is expected since more fitted parameters are used. When all four parameters are fitted together, errors may be distributed among the parameters so as to make extrapolation hazardous.

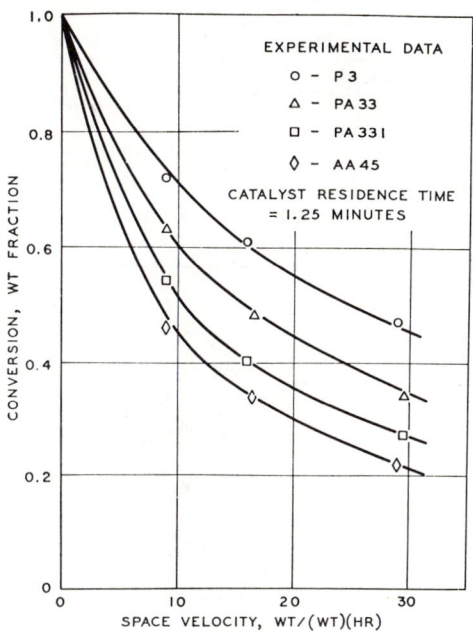

Industrial and Engineering Chemistry,
Process Design and Development

Figure 13. Conversions for different charge stocks (15) (solid line: Equation 7, rate constants Table IV in Ref. 15)

Weekman and Nace (*12*) used a moving bed reactor which enabled them to measure independently the reaction rate and the decay. This paper also presented kinetic equations which allow prediction of the gasoline yield. Figure 13 depicts a typical fit of the model to a wide range of charge stocks. As in the Stangeland model (*10*), the rate constants are strong functions of both the charge stock and the catalyst. Figure 14 is an example of correlation of the rate constants in terms of the aromatic-to-naphthene ratio [Nace *et al.* (*15, 16*)]. Surprisingly, all virgin stocks had a similar correlation whereas materials that had been reacted previously, either by thermal coking or catalytic cracking, differed markedly. This is also illustrated in Figure 15 where either the

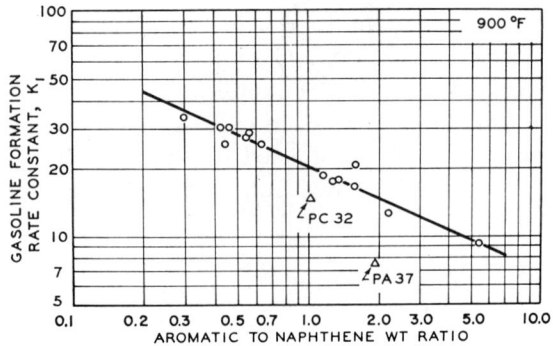

Industrial and Engineering Chemistry,
Process Design and Development

Figure 14. Relationship between gasoline formation rate constant and aromatic-to-naphthene ratio (16)

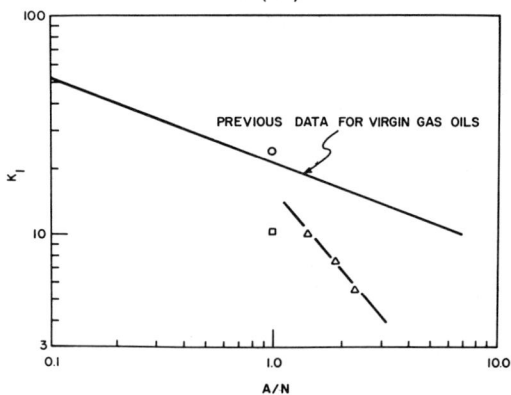

Industrial and Engineering Chemistry,
Process Design and Development

Figure 15. Plot of log K_0 vs. log A/N (17). ○, MCGO; □, MCGO + 0.1 wt % quinoline; and △, FCC fresh feed and combined feeds.

addition of basic nitrogen as quinoline or the addition of recycle material causes a deviation from the virgin gas oil correlation [Voltz et al. (17)]. Since the virgin gas oils were derived from a variety of petroleum sources, the correlation indicates similarity in the formation processes of petroleum from different areas. Thus the aromatic/naphthene ratio is sufficient to pin down the particular distribution of compounds which gives a certain lumped cracking rate. Figure 16 shows that the decay velocity constant also correlates well with the aromatics-to-naphthene ratio.

In the derivation of the kinetic equations, catalyst decay was related solely to catalyst residence time. However, since Voorhies (18) demonstrated that coke formation is strongly dependent on catalyst residence time, it is not at all surprising that a strong correlation exists (*see* Figure 17) between the catalyst decay velocity and the coke on catalyst. This finding contradicts that of Wojciechowski and co-workers (*14*) when they used a much smaller range of charge stock composition than that used by Nace *et al.* (*15*). The Voorhies relationship predicts that coke will be a function only of time on-stream and that it is independent of such parameters as conversion and space velocity. This relationship implies that the charge stock makes coke at about the same rate as the products. In Wojciechowski's recent work, the aromatics were removed

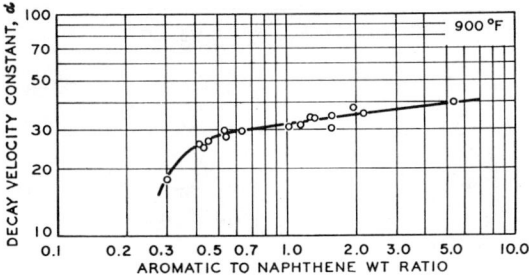

Industrial and Engineering Chemistry,
Process Design and Development

Figure 16. Relationship between catalyst decay constant and aromatic-to-naphthene ratio (16)

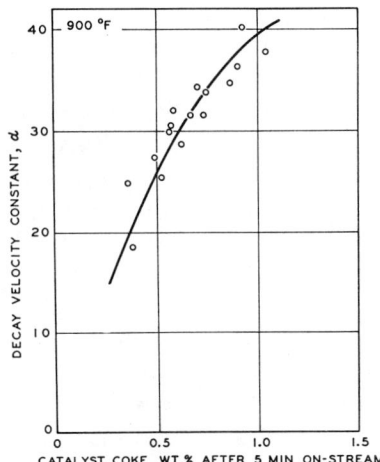

Industrial and Engineering Chemistry,
Process Design and Development

Figure 17. Decay velocity constant vs. catalyst coke (16)

MOLECULAR REACTIONS

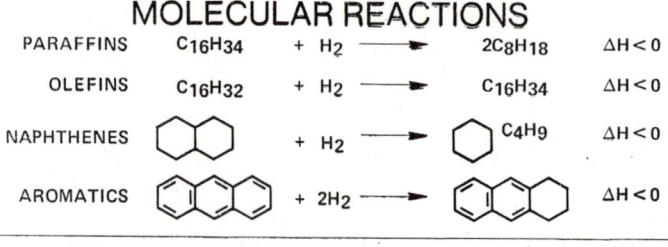

BOND REACTIONS

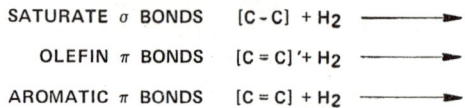

Industrial and Engineering Chemistry,
Process Design and Development

Figure 18. Proposed bond reaction schemes (19)

from the charge stock; consequently coking of charge and of reactants were no longer equal. Under these conditions, a coke profile could be present in the fixed bed, and this could explain the different findings.

Hydrogenolysis Reactions. Jaffe (*19*) recently proposed a novel scheme for lumping hydrogenolysis reactions in order to predict the overall heat release. He noted that, although the heat released by adding hydrogen to olefins, aromatics, cracked paraffins, and naphthenes is quite different, the value within each class is quite similar. In Figure 18 are Jaffe's proposed bond reaction schemes involving saturation of σ bonds and olefin π bonds as well as aromatic π bonds. Figure 19 presents the equivalent kinetic scheme for creating and destroying these bonds. Thus when an aromatic π bond is saturated, σ bonds are created; however, when paraffins crack and subsequently hydrogenate, σ bonds are destroyed. The resulting kinetic equations are listed with the four required rate constants that were determined by fitting from experimental data.

Of course, any lumping scheme must be accompanied by a technique to measure the lumps. Figures 20 and 21 demonstrate that the bond concentrations may be determined directly by standard test procedures. (Ra and RaS represent the aromatic and saturated rings, respectively.) On determining the rate constants, Jaffe found that one set was sufficient to describe three fairly divergent charge stocks. In Figure 22 are plotted the aromatic π bond concentrations as predicted and as observed experimentally. The kinetic scheme adequately describes the experimental behavior over the three-fold range of concentration.

Hydrogen consumption is also predicted directly by the kinetics; predicted and observed values over a very wide range of hydrogen consumption are plotted in Figure 23. This range covers processes from

$$[C=C] + H_2 \underset{k_2}{\overset{k_1}{\rightleftharpoons}} A\,[C-C] \quad \Delta H = -15 \frac{\text{kcal}}{\text{mole } H_2}$$

$$[C=C]' + H_2 \xrightarrow{k_3} [C-C] \quad \Delta H = -30 \frac{\text{kcal}}{\text{mole } H_2}$$

$$[C-C] + H_2 \xrightarrow{k_4} \quad\quad\quad\quad\quad \Delta H = -10 \frac{\text{kcal}}{\text{mole } H_2}$$

$$\frac{d[C=C]}{dt} = -k_1[C=C] + \frac{k_2}{A}[C-C]$$

$$\frac{d[C=C]'}{dt} = -k_3[C=C]'$$

$$\frac{d[C-C]}{dt} = -[k_2 + k_4][C-C] + A\,k_1[C=C] + k_3[C=C]'$$

Industrial and Engineering Chemistry,
Process Design and Development

Figure 19. Kinetic scheme (19)

PARAFFINS

TOTAL CARBON-CARBON BONDS = ½[4C - H]

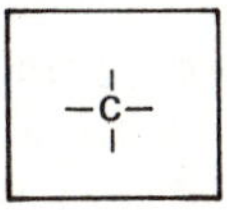

$-\overset{|}{\underset{|}{C}}-$

C-C = ½[4C - H] − 2 C=C′ − 3 C≡C − [Ra - Ras]

Industrial and Engineering Chemistry,
Process Design and Development

Figure 20. Experimental determination of bond concentrations (19)

mild hydrotreating and desulfurization up to deep hydrocracking. As in all the industrial models described so far, these rate constants are strongly dependent on the nature of the catalyst. Although the rate constants seem adequate for describing the three different charge stocks, any large extrapolation beyond these particular stocks could be hazardous.

Theoretical Analysis of Lumping

One of the first comprehensive theoretical analyses of lumped systems was by Wei and Kuo (5, 6). For monomolecular systems they developed a criterion for exact lumpability; that is, they were able to determine which lumps would describe precisely the behavior of the underlying unlumped reaction system. In addition, they described the errors that would result from inexact lumping. Unfortunately, in order

OLEFINS

$$[C=C]' = \frac{\rho}{15.98}\left(\frac{\text{BROMINE}}{\text{NUMBER}}\right) \frac{\text{moles}}{\text{liter}}$$

AROMATICS

$$[C=C] = \tfrac{1}{2}\rho \left[\frac{1-H}{12}\right]\left[\frac{Ca}{C}\right] \frac{\text{moles}}{\text{liter}} \quad \text{n-d-M METHOD}$$

$$[C=C] = \frac{\rho}{MW}\left[\sum_i \tfrac{1}{2} Ca_i\right] \frac{\text{moles}}{\text{liter}} \quad \text{MASS SPECTROMETER}$$

Industrial and Engineering Chemistry, Process Design and Development

Figure 21. Experimental determination of bond concentrations (19)

to utilize the analysis, one must know the underlying system, the avoidance of which is the goal of lumping in the first place. Ozawa (20) recently extended this earlier work and provided criteria for lumping that are based on finding a nonvanishing eigenvector. Luss and Hutchinson (21) described the behavior of lumped systems of parallel first-order reactions, and they analyzed the possible errors in such lumped systems. Golikeri and Luss (22) studied the diffusional problem for lumped systems of parallel first-order reactions. Figure 24 is the result of their analysis for spherical catalyst particles where d represents the dispersion of the lumped system and is the ratio of the variance of the Thiele modulus for the lumped system to the average Thiele modulus.

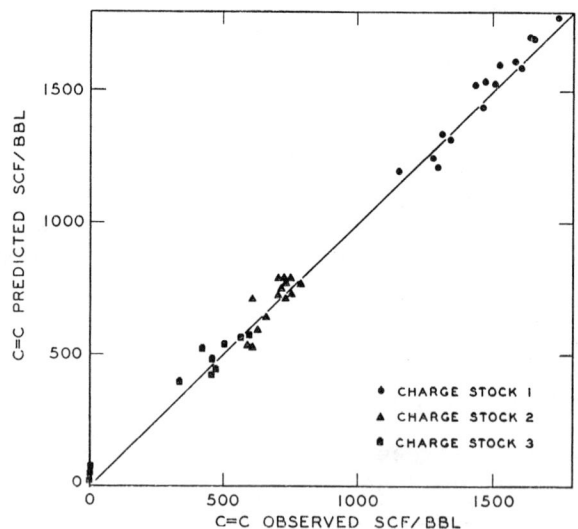

Industrial and Engineering Chemistry, Process Design and Development

Figure 22. Fit to saturation data (19)

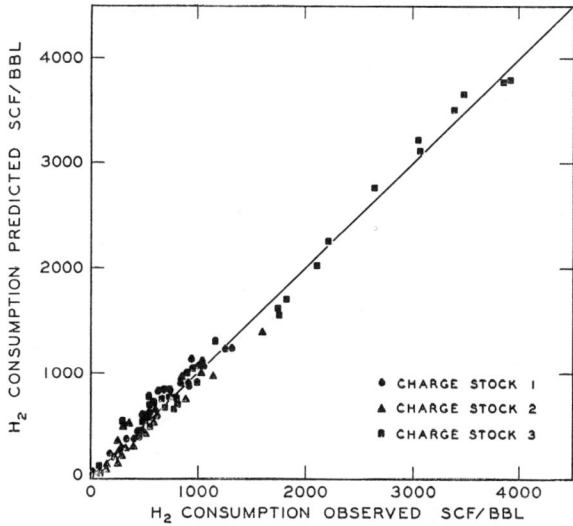

Figure 23. Fit to H_2 consumption data (19)

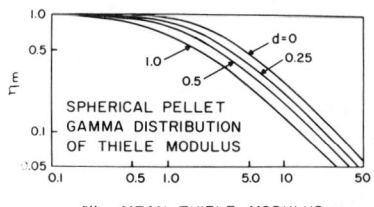

Figure 24. Mean effectiveness factor for first-order reactions inside a spherical pellet (22)

Thus, for a single component or a system where all rate constants are identical, d would be equal to zero.

More recently, Golikeri and Luss (23) represented many coupled, irreversible, consecutive, first-order reactions as a lumped ternary system. They found that the kinetic parameters as well as the functional form of the rate expression may depend on the choice of lumps as well as on the overall feed composition. This demonstrates that one must use great caution when extrapolating lumped systems to feed compositions outside those used in the original study. In an even more frightening example of the problems of lumping, these same authors (24) reported the changes in apparent activation energy which may occur in systems that contain parallel nth order reactions. Figure 25 illustrates the effect on activation

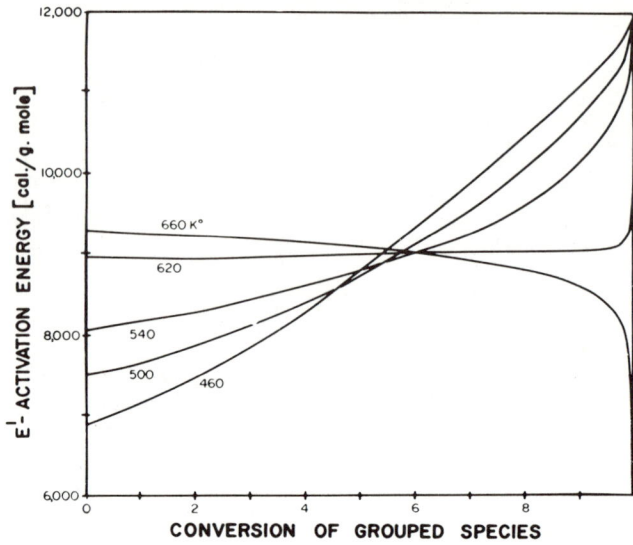

Figure 25. E'—the activation energy of the grouped species at various temperatures and conversion levels (24)

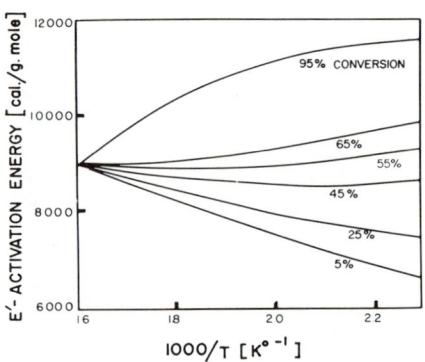

Figure 26. E'—the activation energy of the grouped species at various temperatures and conversion levels (24)

energy of conversion of the lumped species. If we had chosen 620°K as our reaction temperature, we would have found little change in activation energy with conversion and possibly would have been convinced that we had an excellent lumping scheme. However, had we chosen 460°K, a large change in activation energy would have occurred with conversion. Another view of this phenomenon is depicted in Figure 26 where the activation energy is plotted *vs.* temperature. Thus, had we chosen a conversion level of 55% and then varied reaction temperature, we would have found an activation energy which gave a classic Arrhenius plot. However, had we gone to substantially lower or high conversions, we would have found that the activation energy changed substantially with temperature. Again, the lesson is quite clear—that extrapolating lumped systems into regions where data are *not* available can be extremely dangerous. Because significant dangers have already been exposed in this area, additional research work is vital in order to outline further the pitfalls and the degree of error which occur in lumping complex reaction systems.

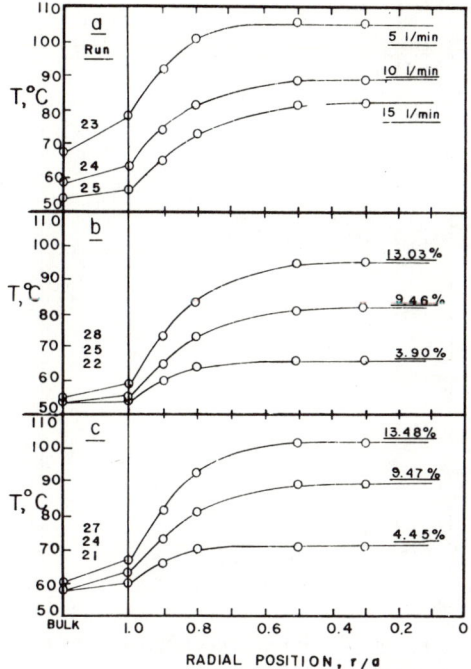

AIChE Journal

Figure 27. Measured internal and external profiles for Pellet 1 with feed temperature of 52°C (25). Profiles as a function of (a) flow rate for feed composition of ~10% C_6H_6; (b) feed composition at high flow, 15 liters/min; and (c) feed composition at intermediate flow, 10 liters/min.

Experimental Confirmation of Internal Heat and Mass Transfer Limitation

In an excellent experimental study, Kehoe and Butt (25) used a carefully instrumented catalyst sphere to study internal heat transfer.

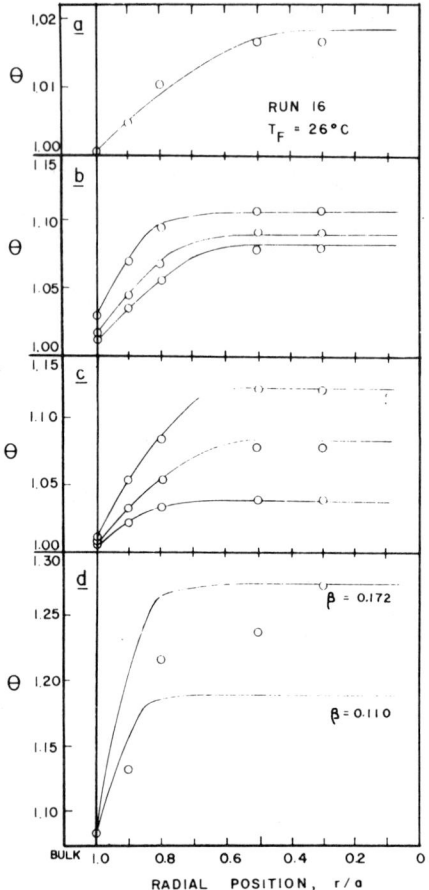

Figure 28. Predicted and observed profiles for Pellet 1 ($T_F = 26°$ and $52°C$) (25). (a) Typical fit to data at $26°C$; (b) fit to feed flow change; (c) fit to feed concentration change; and (d) fit to anomalous profile at high reaction rates.

Some of their measurements for the nickel-catalyzed hydrogenation of benzene are plotted in Figure 27. Not only the internal temperature rise, but also the temperature rise across the film can be observed. Their observed profiles agree well with those predicted by theory (Figure 28). Finally, there was good agreement between their experimentally determined effectiveness factor and that predicted theortically (Figure 29).

Using a similar technique, Koh and Hughes (26) measured internal temperature profiles in a catalyst pellet under fresh conditions and with

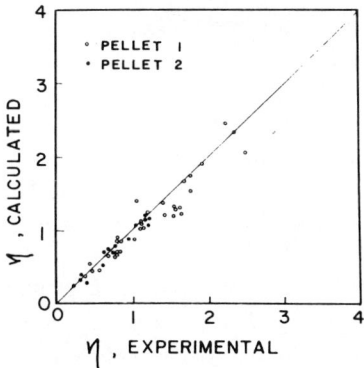

AIChE Journal

Figure 29. Summary of measured and calculated effectiveness factors for all runs (25)

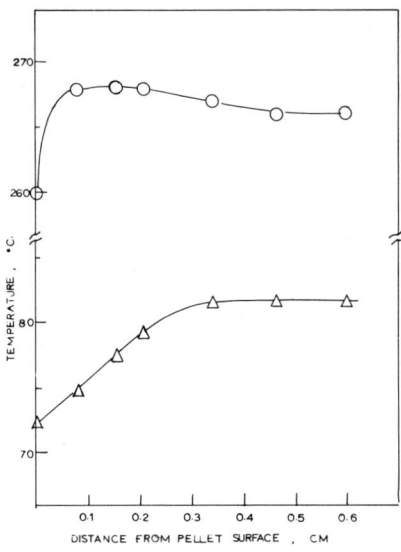

AIChE Journal

Figure 30. Intraparticle temperature profiles for active and poisoned pellet (26). ○: Active pellet, 24% C_2H_4, flow = 21 cm^3/sec, T_{init} = 20°C; △: poisoned pellet, 20.7% C_2H_4, flow = 20 cm^3/sec, T_{init} = 54°C.

a pellet which had been poisoned. The poisoning resulted in a band of nonactive catalyst near the surface (Figure 30). In this region, the temperature rises linearly to the active region inside. The reaction studied was the hydrogenation of ethylene over nickel on silica alumina catalyst; traces of oxygen acted as the poison.

In another experimental study Wang and Wen (27) tested the unreacted shrinking core model where an ash layer provides the controlling mass transfer resistance. In Figure 31 are plotted their measurements that were obtained with fire-clay spheres impregnated with carbon. The unsteady state model appears to give an excellent representation of the experimental data although the pseudo-steady state version is adequate except at the extremes of conversion.

Testing of Fluid Bed Models

An experimental investigation of the merits of existing two-phase fluid bed models was conducted by Shaw *et al.* (28). Their experimental system was the hydrogenolysis of normal butane over a nickel–silica catalyst in an 8-in. diameter fluidized bed pilot plant. The authors listed the models tested (Figure 32) in order of increasing computer time which presumably is close to their ranking in terms of complexity.

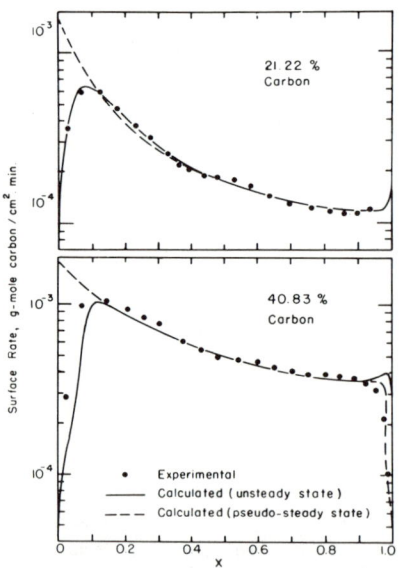

AIChE Journal

Figure 31. Experimental and calculated surface rate–conversion curves (27)

MODEL	UPFLOW PHASE	DENSE PHASE	BUBBLES
ORCUTT ET. AL.	B	PERF. MIX.	D_B CONST.
KATO & WEN	B & C	$U_0 = 0$. DEAD SPACE WITH INTERCHANGE	$D_B = F(H)$
KATO & WEN (MODIF.)	B & C & W		$D_B = F(H)$
ORCUTT ET. AL.	B	PLUG FLOW	D_B CONST.
PARTRIDGE & ROWE	B & C	PLUG FLOW	$D_B = F(H)$
KUNII & LEVENSPIEL	B & C & W	PLUG FLOW UP OR DOWN	D_B CONST.

(COMPUTER TIME increases downward)

Canadian Journal of Chemical Engineering

Figure 32. Summary of existing two-phase models (28). B, bubble void; C, cloud overlap; and W, wake.

A separate fixed bed reactor was used to determine intrinsic kinetic rates, and these values were then applied to all models (*see* Figure 33). All models, with possibly one exception, gave surprisingly similar predictions of conversion as well as selectivity. In a healthy display of candor, the authors then demonstrated that all the models deviated substantially from the experimental results obtained with their pilot plant. Rather than calling down a plague on all the models, they found that one of their rate constants could be adjusted to fit the fluidized bed data. While this brought the models into line with experimental data, it leaves one with an uneasy feeling about the scale-up from the fixed bed to the fluidized bed. Figure 34 is a comparison of the simplest model of Orcutt *et al.* (29) with one of the more complex models proposed by Kato and Wen (30). It is interesting that the more complex bubble cloud and wake model does not appear to be appreciably more accurate than the much simpler Orcutt CSTR model. Such work, studying the

	Bubble Diameter	$D_B = 10.$		
	Gas Interchange	$F_o = 11./D_B$		
	Catalyst Activity	3.6		

Models	Conversion of Butane	Selectivities		
		C_1	C_2	C_3
Orcutt et. al.[7]				
C.S.T.R.	53.3	2.01	0.353	0.427
P.F.T.R.	55.3	2.01	0.352	0.428
C.S.T.R. $U_o = 0$	52.2	2.03	0.356	0.421
Partridge & Rowe[6]	55.5	1.96	0.337	0.457
Kato & Wen[9]	51.0	2.00	0.348	0.435
Wake-Bubble Mixed	57.5	1.93	0.329	-0.471
Assume C.S.T.R. $U_o = 0$				
Wake-Bubble Mixed	53.7	1.85	0.308	0.511
Wake-Emulsion Mixed	52.9	1.95	0.334	0.461
Experimental Results	48.8	2.51	0.302	0.332

Canadian Journal of Chemical Engineering

Figure 33. Comparison of various fluidized bed models (28)

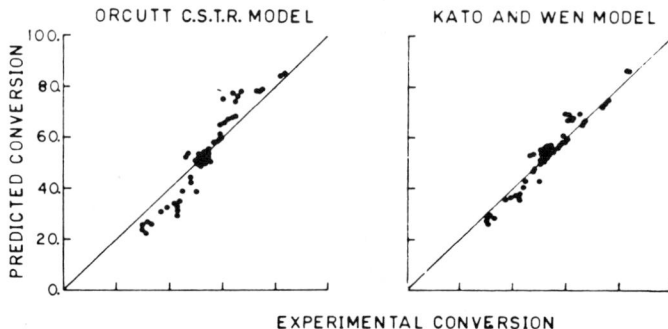

Figure 34. Observed vs. predicted conversion—comparison of the Orcutt model with the Kato and Wen model (28)

effectiveness of various existing models, seems to be a more fruitful exercise than the proliferation of additional models.

Ammonia Synthesis Kinetics

We have long been accustomed to using computers to solve a model that describes a particular reaction. Recently Ferraris *et al.* (*31*) used a computer to generate models which could then be tested against the data. Using the ammonia synthesis reaction, they generated 23 different models for describing the rate of ammonia formation. Of these rate expressions, 21 fit the data almost equally well while one was significantly better than the others and one was significantly worse. Interestingly, the superior rate expression is one which did not appear previously in the literature. Since each rate expression represents an entirely different mechanism, it is again abundantly clear that to determine mechanism from the fit of a rate expression is extremely hazardous. Figures 35 and 36 present examples of some of the rate expressions used by the authors. Since the MSE is the mean squared error with ρ^2 a correlation coefficient and E the average error, the fits for these widely divergent models are essentially equal. It is also clear that each rate expression has significantly different extrapolation properties. Thus, even with the heavily researched ammonia synthesis reaction, we must use care when extrapolating the rate results beyond the bounds of our measurements.

Transport Reactors

A rather heroic example of a process model is that of a commercial kiln for the calcination of ammonium aluminum sulfate to aluminum oxide developed by Manitius *et al.* (*32*). Eleven separate species are

N	Model	MSE	ρ^2	$E\%$	A	B
7	$\dfrac{a_{N_2}a_{H_2} - a_{NH_3}^2/a_{H_2}^2 K^2}{C_1 a_{NH_3}}$	3.93×10^{-4}	0.988	8.46	1.090260	8766.94
8	$\dfrac{a_{N_2}a_{H_2} - a_{NH_3}^2/a_{H_2}^2 K^2}{C_1 a_{N_2} + C_2 a_{NH_3}}$	3.73×10^{-4}	0.988	8.02	−6.539740 0.969128	9667.27 8720.94
9	$\dfrac{a_{N_2}a_{H_2} - a_{NH_3}^2/a_{H_2}^2 K^2}{C_1 + C_2 a_{N_2}^2 + C_3 a_{NH_3}}$	4.01×10^{-4}	0.989	7.82	−13.656600 −6.591480 0.971862	124426.00 9060.79 8770.46
10	$\dfrac{a_{N_2}a_{H_2} - a_{NH_3}^2/a_{H_2}^2 K^2}{C_1 + C_2 a_{NH_3} + C_3 \left[\dfrac{a_{NH_3}}{a_{H_2}}\right]^2 a_{N_2}}$	1.83×10^{-4}	0.995	6.22	1.720570 0.587183 2.450150	8485.22 10287.40 5242.41

Chemical Engineering Science

Figure 35. Models for rate expressions (31)

N	Model	MSE	ρ^2	$E\%$	A	B
11	$\dfrac{a_{N_2}a_{H_2}^3 - a_{NH_3}^2/K^2}{C_1 a_{NH_3}^2 + C_2 a_{N_2} a_{H_2}^2}$	1.81×10^{-4}	0.995	7.42	7.62644 −1.31543	7687.60 10259.80
12	$\dfrac{a_{N_2}a_{H_2}^3 - a_{NH_3}^2/K^2}{C_1 a_{N_2}^2 + C_2 a_{NH_3}^2 + C_3 a_{N_2} a_{H_2}^2}$	1.95×10^{-4}	0.995	7.39	−0.29281 7.63237 −1.32913	5239.81 7752.72 10253.70
13	$\dfrac{a_{N_2}a_{H_2}^3 - a_{NH_3}^2/K^2}{C_1 a_{NH_3}^2 + C_2 a_{H_2} a_{N_2}^2 + C_3 a_{N_2} a_{NH_3}}$	1.93×10^{-4}	0.995	7.57	7.51443 −1.56222 4.23073	7305.49 11165.00 8840.64
14	$\dfrac{a_{N_2}a_{H_2}^3 - a_{NH_3}^2/K^2}{C_1 a_{NH_3}^2 + C_2 a_{N_2} a_{H_2}^2 + C_3 a_{N_2} a_{NH_3}}$	1.95×10^{-4}	0.995	7.40	7.60826 −1.34931 3.44699	7624.06 10229.60 11465.10

Chemical Engineering Science

Figure 36. Models for rate expressions (31)

followed in the kiln along with the related heat balance equations. Figure 37 reveals that the computed gas temperature agrees fairly well with that measured in the commercial kiln. The outer wall temperature, however, did not agree quite as well (Figure 38). While the simulation describes the general shape of the wall temperature, because of the chokes placed on the wall some flow nonuniformity is obviously occurring. The simulation could be checked more accurately if compositional data were available at intermediate points in the kiln to verify mass balance equations. It is always more comforting to check both the heat and the mass balancing equations independently.

The performance of a riser or transport reactor was studied by Pratt (33). This is an increasingly important type in which the catalyst is transported through the reactor by the gaseous reacting system. They are usually large vertical pipes; hence the name riser reactor. Since they are used almost universally in the catalytic cracking of petroleum oils, they represent the largest of all classes of reactors in terms of throughput. Pratt characterized their performance in terms of three dimensionless groups (*see* Figure 39). The M group is the ratio of particle residence time to intraparticle diffusion time and thus represents how many diffusion time constants a given particle spends in the reactor. The P group describes the slip of the catalyst relative to the gas and thus represents a dimensionless catalyst concentration. ϕ is the familiar Thiele modulus

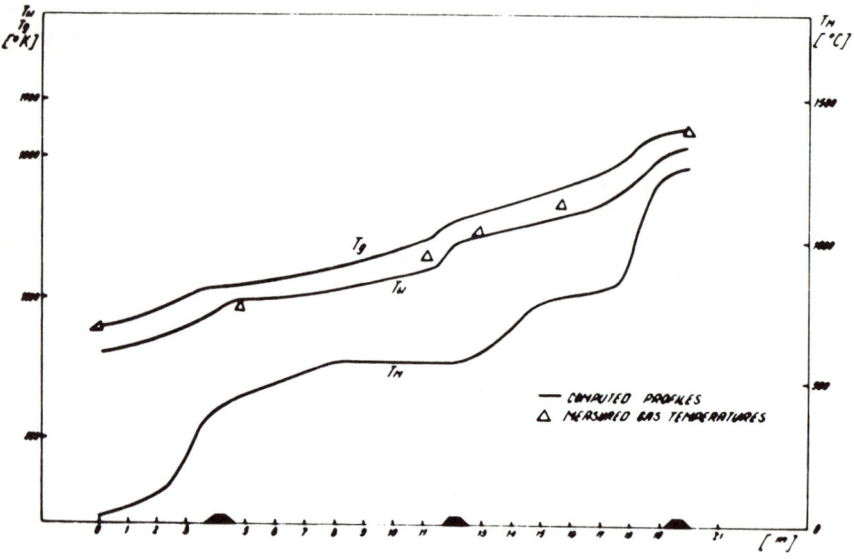

Industrial and Engineering Chemistry, Process Design and Development

Figure 37. Comparison of computed and experimental gas temperature profiles, and computed wall and material temperatures for kiln with chokes (32)

which accounts for diffusion within the catalyst particles. Pratt presented a series of performance charts, one of which is reproduced as Figure 40; conversion increases with increasing M and also with P. Riser reactors are particularly attractive for catalyst systems that use a highly active yet rapidly decaying catalyst. The main difficulty with such systems is the flow patterns that are induced by the separate introduction of catalyst and reactant. The interesting flow patterns that can occur in such reactors were reported earlier by Saxton and Worley (*34*). Much more work is necessary to describe the flow behavior in riser reactors as well as in those that use a higher catalyst concentration (*e.g.* the choked risers or fast fluid beds).

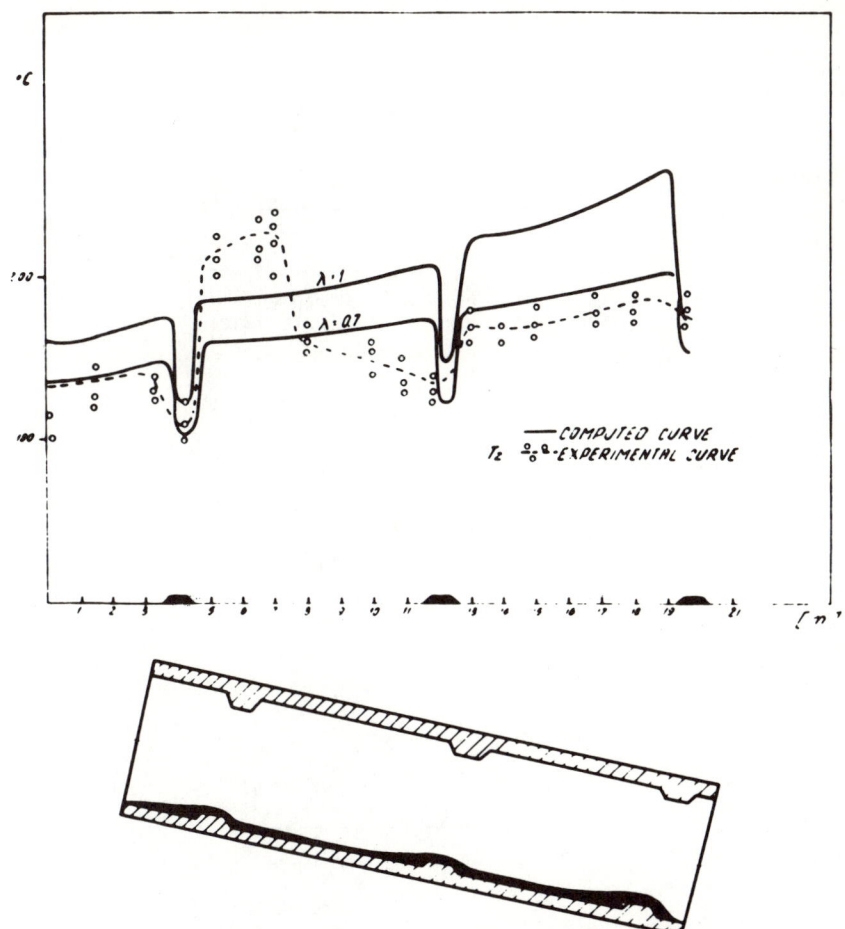

Figure 38. *Process model of a commercial kiln (32). Computed and experimental outer wall temperature profiles (top); and hypothetical shape of the material bed (thin layer, fast motion) in the rotary kiln with chokes (bottom).*

$$M = \frac{DL}{\epsilon R^2 u}, \qquad (19)$$

$$P = \epsilon \frac{u}{U} \frac{(1-\alpha)}{\alpha}, \qquad (20)$$

$$\phi = R\sqrt{\frac{k}{D}}. \qquad (21)$$

Figure 39. *Three dimensionless groups that characterize the performance of a riser reactor (33)*

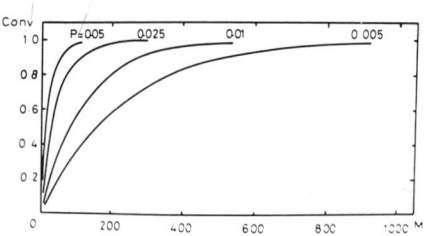

Chemical Engineering Science

Figure 40. Conversion as a function of the parameter M for various values of P with $\phi^2 = 1$ (33)

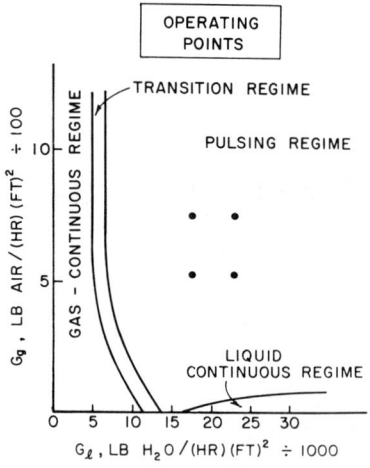

AIChE Journal

Figure 41. Flow regimes observed by Wulfert (36)

Gas–Liquid Flow in Packed Beds

A thorough review of trickle bed reactors is presented in this volume by Satterfield (35); therefore, this review will concentrate on flow rates higher than those used in trickle flow. Beimesch and Kessler (36) mapped the various regimes of flow observed in past work. In Figure 41, the major flow regimes are plotted in terms of the liquid and gas loadings in packed reactors; the pulsing regime is dominant at higher gas and liquid rates. More recently, Sato et al. (37) published a slightly more detailed map of observed flow regimes (Figure 42). The primary difference between the two diagrams is the relabeling of the liquid continuous regime as dispersed bubble flow.

Of key importance is the transition from the gas continuous to the pulsing or liquid continuous regime. Such transition can occur when a laboratory reactor is scaled up to pilot plant or commercial size. As we increase the length-to-diameter ratio, as is typical with the larger reactors, the relative liquid and gas loadings increase at constant space velocity. Thus, whereas the laboratory reactor may be in the gas continuous region, the pilot plant or commercial reactor may be in the pulsing regime. Unfortunately, very few, if any, kinetic studies have been made across this transitional interface. Earlier work by this reviewer (38) demonstrated that the heat transfer in the radial direction can change greatly when the flow changes from the gas continuous to the pulsing regime. This greatly enhances radial heat transfer and may also imply improved contacting with significant effects on the kinetics.

Beimesch and Kessler (36) investigated the structure of the pulses; their proposed pulsing flow model is depicted in Figure 43. Here the dominant feature is a slug of liquid followed by gas continuous regions.

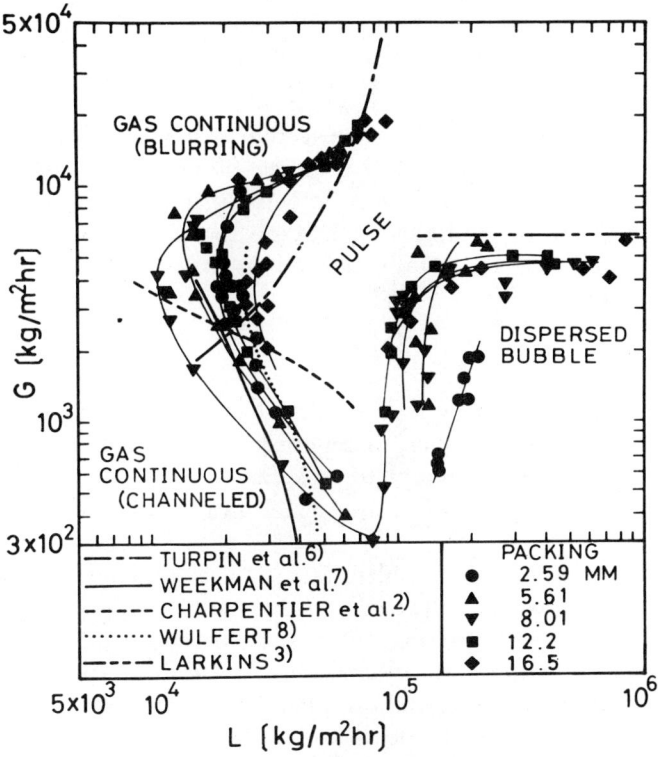

Journal of Chemical Engineering of Japan

Figure 42. Summary of present and published diagrams for flow pattern boundary (37)

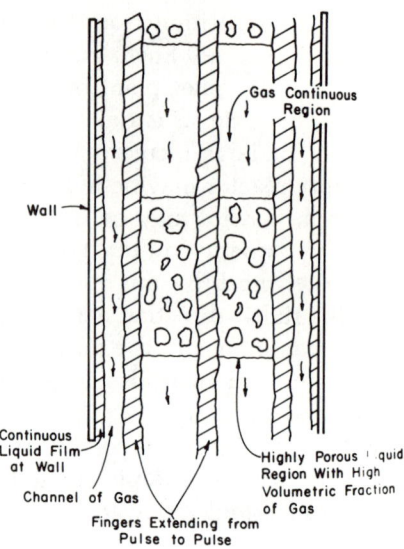

AIChE Journal

Figure 43. Proposed pulsing flow model (36)

They propose a channel of gas near the wall with liquid fingers that extend between the pulses. The presence of such pulses alters the contacting efficiency of the bed with potentially significantly different mass transfer effects in the pulses and between the pulses. Kinetic studies in this regime of flow are vitally needed in order to determine realistic design criteria.

Summary

While much progress is being made toward solving the problems faced by the reactor designer, many areas require further research work. This reviewer will not attempt to make an exhaustive summary of these areas but rather will identify those which in his opinion are particularly ripe.

Probably the greatest problem facing those who apply the principles of reaction kinetics to industrial reactors is the fact that all the rate constants are strong functions of the particular catalyst used. Thus small changes in catalyst composition, or changes induced in the catalyst by the aging process, necessitate re-identification of all the rate constants that were used in the kinetic model. At the present time, very few empirical correlations exist to guide the designer in predicting the effect of changes in catalyst properties on the resultant kinetic rate constants. We have not yet even developed our J-factor analogy in this area. Solution of this problem requires very close cooperation between those involved in reaction engineering and those involved in surface chemistry. Any theories, correlations, or even guidelines in this area will be very valuable since the greatest research expenditure in the process industry today is probably for the repeat experimental work that is necessitated by changes in catalyst properties.

Important progress has been made in studies of lumped reaction systems; they have revealed primarily some of the possible pitfalls when such lumped systems are used. The major problem is to identify the best lumps to choose before one has analyzed the system exhaustively. Suitable guidelines can greatly reduce the amount of experimental work required in determining the lumpability of complex systems. Such work will, of course, be closely connected with the analytical procedures required to identify the correct lumps in the reacting mixtures. Without such guidelines, experimental programs to re-analyze the rate constant behavior of each new feedstock would be extremely expensive.

Th greatest difficulty encountered in scaling up reactors to commercial size is probably that in predicting the flow patterns in the large scale reactor. Unusual flow patterns may be induced in large fluid bed reactors by reactor internals while riser reactors may have uneven catalyst–reactant profiles that are caused by initial mixing phenomena. In between the velocities of the riser and the fluid bed, the fast fluidized bed or choked riser reactors may exhibit complex flow behavior in both the catalyst and the reactant phases. As was described earlier, the flow patterns in fixed bed reactors undergoing two-phase flow may be exceedingly complex. The transition from one flow regime to another may cause significant changes in reactor performance. Such fluid dynamic and mixing behavior may, in fact, dominate the behavior of the scaled-up industrial reactor.

It is also well to keep in mind that, unless rate constants are determined under essentially gradientless conditions in the laboratory, they may be contaminated with mixing, heat and mass transfer, or fluid dynamic effects present in the laboratory reactor. It is very unlikely that such effects will scale up properly to the commercial design. Some recent reviews addressed this problem of designing laboratory reactors to determine industrial rate constants (*39, 40*). In the area of mass transfer limitations, the presence of liquid in the catalyst pores may complicate reaction behavior, particularly if noncondensable gases are produced by the reaction. Study of two-phase flow in catalyst pores is an area in which little research work has been done.

Finally, it is vitally important that industry release more well documented reactor performance data. Such data can serve to test existing theories and possibly highlight new important research areas. It is very important, however, that the data be well documented since the interpretation of commercial data can be quite hazardous because of noise or changes in certain key parameters which were not controlled. Data on obsolete processes could be published since the competitive edge has been lost, yet the data can still be important for providing tests of theory or clarifying reactor performance phenomena.

Literature Cited

1. Prater, C. D., private communication (1970).
2. Prater, C. D., private communication (1972).
3. Aris, R., Gavalas, G. R., *Philos. Trans. R. Soc. London* (1966) **260**, 351.
4. Aris, R., *Arch. Ration Mech. Anal.* (1968) **27**, 356.
5. Wei, J., Kuo, J. C. W., *Ind. Eng. Chem. Fundam.* (1969) **8**, 114.
6. *Ibid.* (1969) **8**, 124.
7. Smith, R. B., *Chem. Eng. Prog.* (1959) **55** (6), 76.
8. Dorokhov, A. P., Ioffe, I. I., Maslyanskii, G. N., Federov, A. P., Fuks, I. S., Shipikin, V. V., *Khim. Tekhnol. Topl. Masel* (1972) **11**, 6.
9. Kmak, W. S., "A Kinetic Simulation Model of the Powerforming Process," AIChE National Meeting, Houston, Tex. (1971).
10. Stangeland, B. E., *Ind. Eng. Chem. Process Des. Dev.* (1974) **13**, 71.
11. Weekman, Jr., V. W., *Ind. Eng. Chem. Process Des. Dev.* (1968) **1**, 90.
12. Weekman, Jr., V. W., Nace, D. M., *AIChE J.* (1970) **16**, 397.
13. Wojciechowski, B. W., *Can. J. Chem. Eng.* (1968) **46**, 48.
14. John, T. M., Pachovsky, R. A., Wojciechowski, B. W., Adv. Chem. Ser. (1974) **133**, 422.
15. Nace, D. M., Voltz, S. E., Weekman, Jr., V. W., *Ind. Eng. Chem. Process Des. Dev.* (1971) **10**, 530.
16. *Ibid.* (1971) **10**, 538.
17. Voltz, S. E., Nace, D. M., Jacob, S. M., Weekman, Jr., V. W., *Ind. Eng. Chem. Process Des. Dev.* (1972) **11**, 261.
18. Voorhies, A., *Ind. Eng. Chem.* (1945) **37**, 318.
19. Jaffe, S. B., *Ind. Eng. Chem. Process Des. Dev.* (1974) **13**, 34.
20. Ozawa, Y., *Ind. Eng. Chem. Fundam.* (1973) **12**, 191.
21. Luss, D., Hutchinson, P., *Chem. Eng. J.* (1971) **2**, 172.
22. Golikeri, S. V., Luss, D., *Chem. Eng. Sci.* (1971) **26**, 237.
23. *Ibid.* (1974) **29**, 845.
24. Golikeri, S. V., Luss, D., *AIChE J.* (1972) **18**, 227.
25. Kehoe, J. P. G., Butt, J. B., *AIChE J.* (1972) **18**, 347.
26. Koh, H.-P., Hughes, R., *AIChE J.* (1974) **20**, 395.
27. Wang, S. C., Wen, C. Y., *AIChE J.* (1972) **18**, 1231.
28. Shaw, I. D., Hoffman, T. W., Orlickas, A., Reilly, P. M., *Can. J. Chem. Eng.* (1972) **50**, 637.
29. Orcutt, J. C., Davidson, J. E., Pigford, R. L., *Chem. Eng. Prog. Symp. Ser.* (1962) **58** (38), 1–15.
30. Kato, K., Wen, C. Y., *Chem. Eng. Sci.* (1969) **24** (8), 1351–1369.
31. Ferraris, G. B., Donati, G., Rejna, F., Carra, S., *Chem. Eng. Sci.* (1974) **29**, 1621.
32. Manitius, A., Kurcyusz, E., Kawecki, W., *Ind. Eng. Chem. Process Des. Dev.* (1974) **13**, 132.
33. Pratt, K. C., *Chem. Eng. Sci.* (1974) **29**, 747.
34. Saxton, A. L., Worley, A. C., *Oil Gas J.* (May 18, 1970) **68** (20).
35. Satterfield, C. N., "Contacting Effectiveness in Trickle Bed Reactors," Adv. Chem. Ser. (1975) **148**, 50.

36. Beimesch, W. E., Kessler, D. P., *AIChE J.* (1971) **17**, 1160.
37. Sato, Y., Hirose, T., Takahashi, F., Toda, M., Hashiguchi, Y., *J. Chem. Eng. Jpn.* (1973) **6**, 315.
38. Weekman, Jr., V. W., *AIChE J.* (1965) **11**, 13.
39. Doraiswamy, L. K., Tajbl, D. G., *Catal. Rev., Sci. Eng.* (1974) **10**, 177.
40. Weekman, Jr., V. W., *AIChE J.* (1974) **20**, 833.

RECEIVED December 4, 1974.

6

The Role of Chemical Reaction Engineering in Coal Gasification

HERMAN F. FELDMANN

Battelle, Columbus Laboratories, 505 King Ave., Columbus, Ohio 43201

The stoichiometry and the unit operations required to convert coal to a clean gaseous fuel are briefly described. The most important reactions occurring in this conversion are pyrolysis, hydrogasification, and steam gasification. Gross physical transformations occurring during these reactions are discussed as well as what appears to be a reasonable qualitative basis for the development of reaction rate equations describing each step. Rate equations developed by various workers are described, and the physical bases for these equations are discussed. Further experimental work should focus on establishing more accurately what is occurring in the complex system than on generating additional reaction rate equations. Gasifier designs proposed to optimize the complex overall gasification system are described. Since most of these reactor systems use relatively complex gas–solid flow circuits, recommendations are made for increased experimental and analytical studies to simplify the design and operational control of these reactor systems.

Coal can react with steam, carbon dioxide, or hydrogen to produce gaseous products that can be used either directly for fuel or for synthesizing a variety of other chemical compounds. This conversion involves the three main reactions listed below:

$$C + H_2O \rightleftharpoons CO + H_2 \quad \text{(Gasification)} \tag{1}$$

$$C + CO_2 \rightleftharpoons 2CO \quad \text{(Gasification)} \tag{2}$$

$$C + 2H_2 \rightleftharpoons CH_4 \quad \text{(Hydrogasification)} \tag{3}$$

In addition, there are two gas phase reactions important at gasification conditions—the so-called water–gas shift reaction (Reaction 4) and the methanation reaction (Reaction 5):

$$CO + H_2O \rightleftarrows CO_2 + H_2 \qquad (4)$$

$$CO + 3H_2 \rightleftarrows CH_4 + H_2O \qquad (5)$$

The reverse of Reaction 5 is called methane–steam reforming and is undesirable when methane is the desired product.

Thus, one can start with a carbonaceous solid such as coal, and after gasification and hydrogasification one ends with CH_4, CO, and H_2 rather than a dirty solid. The CO and H_2 may react together further to form additional paraffinic hydrocarbons ranging from methane to gasoline and waxes as well as alcohols, olefins, and in short, almost any compound that can be formed from petroleum as well as a synthetic natural gas (SNG). The ability of gasification to transform the carbon in coal to compounds that can supplement our dwindling supplies of petroleum and natural gas combined with a domestic supply of coal that can satisfy our energy needs for hundreds of years has given the development of coal gasification technology an extremely high national priority.

The greatest emphasis in the U.S. is to gasify coal to produce the following primary gaseous products:

(1) Substitute natural gas (SNG) to supplement rapidly diminishing natural gas. Heating value of SNG will vary from about 900 to 980 Btu/SCF.

(2) Synthesis gas, which consists mainly of CO and H_2 with some CH_4, has a heating value of about 300 to 500 Btu/SCF depending on the methane concentration. Syngas can be utilized in utility boilers in place of natural gas or for synthesizing a variety of chemicals ranging from paraffinic waxes to gasoline, olefins, and alcohols.

(3) Producer gas which has a heating value of about 100 to 150 Btu/SCF because of the nitrogen diluent introduced by using air as a gasifying agent. This gas could be used in utility or industrial boilers.

If economic technologies to produce the above products from coal can be developed, the United States will be able to bring its energy consumption patterns more in proportion to its reserves of coal, oil, and natural gas. For example, in 1972 petroleum supplied about 46% of all domestic energy requirements followed by natural gas (32%), bituminous coal (17%), hydropower (4%), nuclear (0.8%), and anthracite (0.2%). To supply this energy demand pattern, the United States, formerly a fuel exporter, had to rely ever increasingly on imported fuels. In 1972, for example, we imported about 12.5% of our fuel which represented a 24.5% increase over 1971.

Generic Process Descriptions

In this section we briefly and broadly discuss the integrated unit operations necessary to convert coal to gas. Detailed process descriptions are available from many sources. [Excellent sources are the Office of Coal Research reports, preprints of the ACS Division of Fuel Chemistry, and the Proceedings of the AGA/OCR Synthetic Pipeline Gas Symposia.] Therefore, only a generic description of each process type is given to point out the differences in basic types of processes and to show how the gasification operation relates to the other unit operations necessary to produce the final product.

SNG and Synthesis Gas. A typical process to produce SNG from coal is shown in Figure 1. The first major processing step is pretreatment to prevent coal swelling and agglomeration which would cause plugging and shutdown of the gasification reaction system. This step is necessary only for Eastern caking coals and can be eliminated for Western sub-bituminous coal and lignite. Pretreatment is simply the partial oxidation of the volatile constituents of the coal which would otherwise cause it to soften, swell, and stick. Removal of this volatile matter is economically and technically undesirable because its loss reduces coal reactivity and methane yield and increases oxygen consumption. Pretreatment is

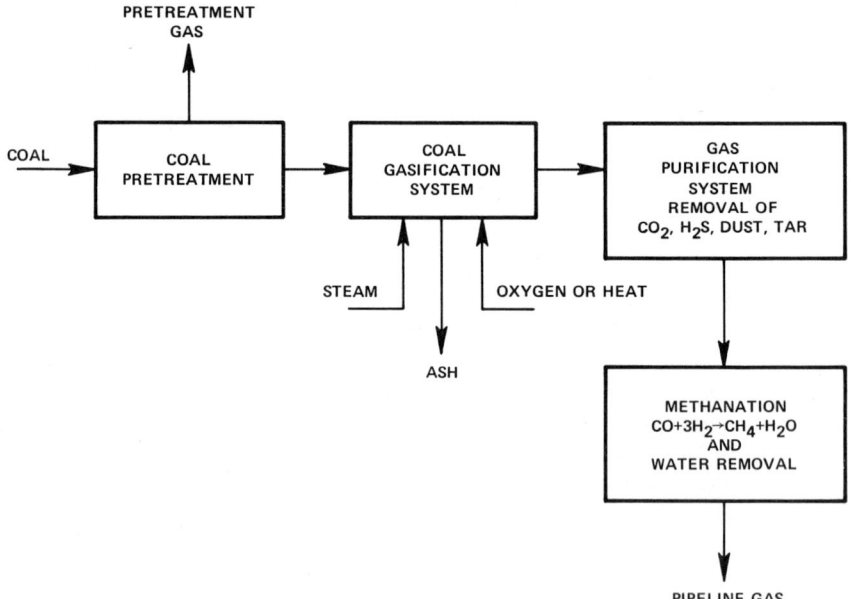

Figure 1. *Simplified flow diagram for producing pipeline gas from coal*

avoided in the Bureau of Mines Hydrane Process and Bituminous Coal Research's Bi-Gas Process by utilizing dilute phase reactors.

After pretreatment the coal goes to the (hydro)gasification reactor system which is the heart of the process. In the reactor system, which can be quite complex, stoichiometric Reactions 1 through 4 occur except that coal or char is the solid reactant instead of carbon which greatly increases the kinetic and thermodynamic complexity. [Char is defined as the solid carbonaceous product arising from the thermal treatment of coal.] An additional, extremely important reaction occurs, the so-called pyrolysis or devolatilization reaction.

$$\text{Coal} + \text{heat} \rightarrow \text{gases } (CH_4, CO, CO_2, H_2, H_2O) + \text{tar} + \text{char} \quad (6)$$

Gasification reactor systems used for SNG production all contain two basic elements or zones. In the first zone the incoming coal first passes an environment in which conditions are controlled as nearly as possible to maximize the yield of methane by Reaction 3, the hydrogasification reactions, and Reaction 6, the devolatilization reaction. To maximize methane production, favored at elevated pressure, and to avoid compressing the gas for the pipeline, the optimum pressure is on the order of 1000 psig (69 atm). Hydrogasification temperatures range from 1200° to 1800°F (650°–980°C). At temperatures much higher than 1800°F, methane begins to decompose. The hydrogen for hydrogasification is produced in the second zone from the so-called gasification reactions (Reactions 1 and 2), and the water gas shift reaction (Reaction 4). Gasification reactions are not especially sensitive to pressure but require temperatures higher than that required for optimal methane formation. Typical temperature ranges are from 1800°F (980°C) up to 2800°F (1540°C) for slagging gasification. For this reason the reactor must be separated into zones. The heat for the highly endothermic gasification reactions occurring in the second zone is provided by one of the following means:

(a) Burning part of the carbon with oxygen, which is the technique that is most technically advanced.

(b) Burning part of the carbon with air to provide heat to an intermediate heat transfer agent such as agglomerated ash (1) or calcined dolomite (2). These intermediate heat transfer techniques allow the combustion heat to be transferred to the gasification system without diluting the synthesis gas with nitrogen from the air. There is a large economic potential in such systems because they allow nitrogen-free synthesis gas to be generated without an expensive oxygen plant. The technical problem with such a system is the requirement to circulate a large amount of the heat transfer agent to satisfy the endothermic heat requirements of the gasification reaction at the high temperatures required to achieve reasonable gasification rates.

(c) In the IGT steam–iron process (3) a producer gas is generated which is used to reduce iron oxides. The reduced iron–iron oxide then

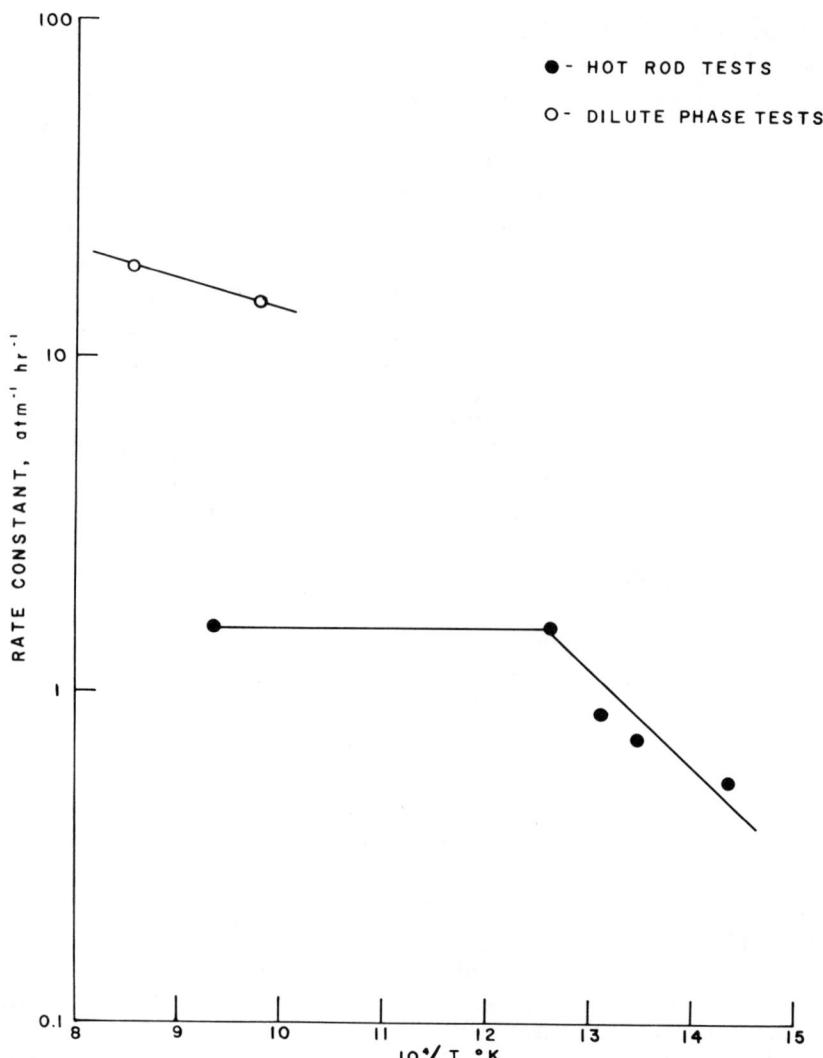

Figure 2. Effect of temperature and reactor type on hydrogasification rate (data courtesy Pittsburgh Energy Research Center—ERDA)

reacts with steam to produce a hydrogen–steam mixture for the hydrogasification of coal to a methane-rich gas.

Because the conditions for producing methane are not suitable for generating synthesis gas and vice-versa, most reactor systems for producing SNG are, as we shall see, rather complex.

If the gasification system is intended to produce synthesis gas rather than methane, considerable reactor simplifications may be made. For example, a single-zone pressurized entrained gasifier can be utilized instead of the multi-zoned reactor systems necessary to maximize methane production in the reactor system. Since methane can be synthesized from synthesis gas by the well-known methanation reaction, a logical question is: why use the more complex reactor systems needed to maximize direct production of methane from coal?

The answer is that the direct formation of methane results in improved thermal efficiency (4, 5, 6). In addition, other process simplifications (7) result from maximizing the fraction of methane produced by hydrogasification that can substantially lower costs for the production of SNG (5). However, the reliability of the more complex reactor systems has not yet been established. Therefore, before the "best" system can be selected, reactors of at least pilot-plant scale must be operated for sufficiently long times to allow reliability to be accurately established.

Some other criteria that are important in selecting the "best" reactor system for either the production of synthesis gas or SNG will be:

(a) The ability of the reactor system to handle a wide range of coal particle sizes. For example, continuous mining machinery produces a large fraction of coal too fine for the conventional Lurgi (8) moving bed reactor system, though experiments are in progress at Westfield, Scotland to modify the Lurgi to handle finer coal (9, 10).

(b) The ability to handle highly caking Eastern U.S. coals without pretreatment.

(c) The ability to be scaled up large enough to minimize the number of reactor trains required for a commercial SNG plant. For example, if the Lurgi system is used, about 30 reactors will be necessary for a plant producing 250 million SCF/day of pipeline quality gas.

Gas Purification. The hot raw gas from the gasification system contains the following components in amounts that depend on the specific kind of gasification system used:

CH_4	H_2S
CO	COS
H_2	NH_3
CO_2	dust
H_2O	tar

Basic steps in purifying the gas include:

(1) Cooling the gases and the removal of dust and tar.

(2) Water–gas shift ($CO + H_2O \rightleftarrows CO_2 + H_2$) to adjust the H_2/CO ratio to the value desired for the particular synthesis job required. For example, if CH_4 is the desired product, the H_2/CO ratio would be adjusted to slightly over 3; if, on the other hand, hydrogen were the desired product, the shift reaction would be carried to completion.

(3) Removal of acid gases, H_2S, COS, and CO_2. Several commercial processes are available for removal of these constituents from the cleaned coal gas.

(4) Methanation (Reaction 5) or other synthesis reactions to produce a variety of hydrocarbon products from CO and H_2.

For the most part commercial technology is available for performing the above purification steps.

Low-Btu Gas (Producer Gas). If the combustion of carbon with air rather than oxygen directly provides the heat for the carbon–steam reaction, the gaseous product will contain nitrogen in addition to the species previously listed as being in the oxygen blown gasification systems. Ordinarily the nitrogen content of low-Btu gas is approximately 50%, and the heating value will vary from about 100 to 150 Btu/SCF.

Since the manufacture of synthesis gas does not require oxygen, it will be in most cases cheaper on a Btu basis than synthesis gas. However, if much downstream treatment, such as cooling and purification, is necessary, increased capital and operating costs caused by the necessity of handling greater volumes of nitrogen-diluted gas can quickly negate the savings gained by the elimination of an oxygen plant, especially for large installations.

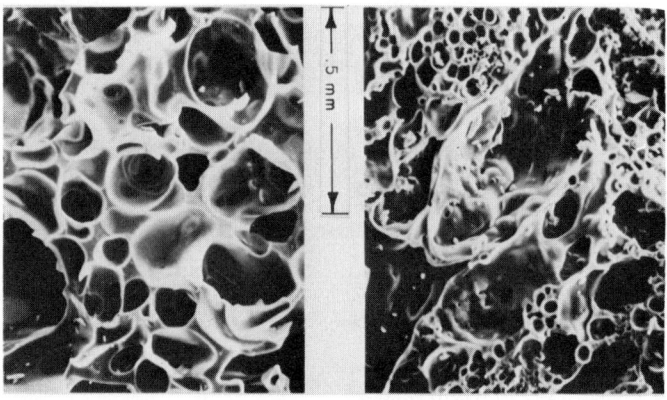

Figure 3. Char sections after hydrogasification, ×48 (courtesy Pittsburgh Energy Research Center—ERDA)

Also, because of its low volumetric heating value, low-Btu gas cannot be transmitted very far. Thus, its major uses are expected to be for on-site utilization for certain industrial applications as a substitute for natural gas, or as a fuel for either conventional utility plants or combined cycle power plants.

There are two major advantages in converting coal to low-Btu gas for combustion purposes rather than burning the coal directly and removing the sulfur by stack gas scrubbing: (1) the combustion of gas eliminates or greatly reduces particulate emissions; and (2) in low-Btu gas the sulfur is present mainly as H_2S which can be removed using available H_2S

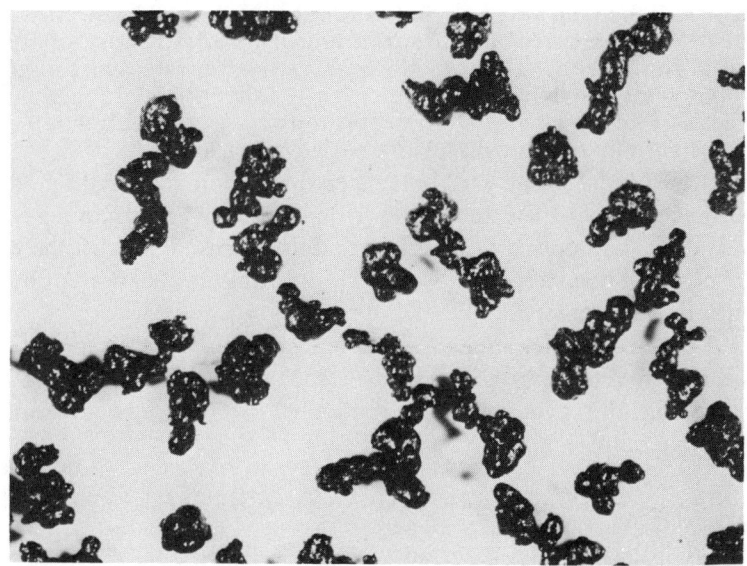

Figure 4. Char particles after dilute-phase hydrogasification (courtesy Pittsburgh Energy Research Center—ERDA)

removal technology, converting the coal sulfur to high-grade elemental sulfur rather than the sulfate sludge product of stack gas scrubbing.

In principle, any gasification system used to produce synthesis gas or SNG can also be used to produce low-Btu gas by simply substituting air for oxygen. [In practice there are exceptions to this statement. For example, certain slagging bed gasification systems such as the Koppers-Totzek gasifier require such high gasification temperatures for smooth operation that substitution of air for oxygen necessitates extremely high air preheat temperatures.]

Gasification Reaction Rates

The coal or coal char analogs of Reactions 1, 2, and 3 are the reactions of greatest importance in the design and optimization of coal gasification systems. Before discussing gasification kinetics, it is important to recognize two important properties of coal/coal–char reactions:

(1) The reactions of coal char with hydrogen, steam, and carbon dioxide are irreversible. For example, the reaction:

$$\text{coal} + H_2 \rightarrow \text{char} + CH_4$$

cannot be reversed to allow coal to be formed from char and methane. This means that the very handy concept of equilibrium approach cannot be correctly used in analyzing coal reactions.

(2) The reaction rates of coal are highly dependent not only on the type of coal but unfortunately on the environmental history of the coal particle in the reactor. Thus, such things as heatup rate and the gaseous atmosphere and its change with time greatly influence coal reaction rates. This means that in kinetic studies one must pay great attention to the conditions anticipated to exist in the scaled-up reactor.

Because of the great variation of coal reaction rates with both coal type and particle history, this section does not cover this topic exhaustively. Rather, a sampling of reaction models is presented, and the chemical and physical transformations occurring during gasification and hydrogasification are discussed.

Coal–Hydrogen Reactions. The greatest single effort in coal gasification is directed at the production of methane, and the reactions of greatest importance in the production of methane are the devolatilization or pyrolysis reaction

$$\text{coal} \rightarrow \text{char} + CH_4 + CO + H_2 + \text{tar}$$

and the hydrogasification reaction

$$\text{coal} + xH_2 \rightarrow CH_4 + \text{char},$$

where x depends on the type of coal, the hydrogen partial pressure, and the degree of carbon conversion.

Before discussing reaction models in any detail, it is worthwhile to describe the generally accepted transformations that occur while coal is heated up to reaction temperature ($\sim 1000°F$). If an Eastern bituminous coal is used, the coal will start to soften at about 500°F (260°C), and at about 750°F (400°C) the evolution of volatile matter begins (*11*). [Volatile matter is defined as the amount of material that is volatized during heatup of the coal particle. It consists of mostly tar vapors, CH_4, CO, H_2, and bound water.] The vapors released by the thermal decomposition of coal cause the particle to swell and finally to assume an open porous structure after the particle solidifies (*see* Figure 3). The resolidification occurs as a result of the conversion of liquid-forming materials into gases, vapors, and solids. The amount of porosity depends upon many factors, especially the

> type of coal
> heating rate
> gaseous environment
> pressure
> physical constraints against expansion such as reactor walls or other particles.

An explanation for the fact that the rate of carbon hydrogasification (*12*) and pyrolysis (*11*) increase with increasing heatup rate is that the volatized species escape the particle and mix with the hydrogen at a higher rate and thereby decrease the amount of recombination that occurs in the particle.

In most reaction systems of commercial interest the particle heatup rate to the temperatures required for the above transformation is extremely high, occurring in fractions of a second. Therefore, the history of a coal particle during a time period lasting less than a second has a great effect on its subsequent reactivity in the reactor.

The chemical transformations that occur during this period are probably more important than the physical transformations. During this time of softening, swelling, and devolatilization, bonds within the coal molecule are breaking, resulting in the formation of radicals which makes the coal extremely reactive. The reactive species formed can either react together to form highly condensed and thereby unreactive tars or solid residual carbon, or they can react with hydrogen to form hydrocarbon gases, lighter aromatics, or solid reactive species that are in turn further hydrogasified to hydrocarbon gases. The solid carbon structure formed by the condensed radicals can be hydrogenated further but at a much lower rate than either the volatile matter or the reactive solid carbonaceous material.

Table I. High-Btu Gas Processes

Process	*Developer*	*Sponsor*
Bi-Gas	Bituminous Coal Research	AGA/OCR
Hy-Gas	Institute of Gas Technology	AGA/OCR
CO_2 Acceptor	Conoco Coal Development Corp.	AGA/OCR
Battelle/Union Carbide Agglomerating Gasifier	Battelle Institute	
Synthane	U.S. Bureau of Mines	U.S. Bureau of Mines
Hydrane	U.S. Bureau of Mines	U.S. Bureau of Mines

The above picture may be summarized by assuming that the carbon in the coal is distributed into the following phases each having a different reactivity with hydrogen:

(1) The most reactive carbon which is quickly (almost instantaneously) converted to methane. This carbon is contained in aliphatic chains bridging the aromatic groups in the coal. The conversion of this carbon is probably limited by heat transfer.

(2) Carbon that is associated with reactive solids which are probably formed by the reaction of radicals with hydrogen. Though this carbon reacts with hydrogen at a slower rate than does the carbon in the aliphatic chains, its half-life is still only on the order of seconds.

(3) Residual carbon formed by the condensation of radicals and which reacts with hydrogen with a half-life ranging from minutes to hours.

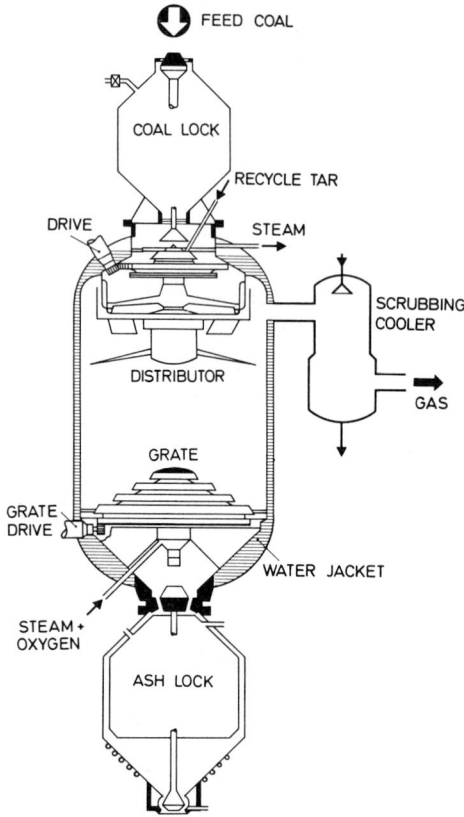

Figure 5. Lurgi pressure gasifier

Although they may differ in the manner in which the mathematical forms for the rate equations are developed, most of the reaction models proposed for the reaction of coal or char with hydrogen to form methane are based on the above assumptions.

The fraction of total carbon in the coal that can be assigned to each of the above three categories depends on the conditions that the coal particle sees upon entering the reactor. For example, Mosely and Patterson (*13*) achieved complete conversion with residence times on the order of 1 sec or less when hydrogen partial pressures approached 1000 atm. Feldmann *et al.* (*14*) with data generated by Hitshue *et al.* (*15*) determined that the fraction of Phase 1 and Phase 2 carbon is determined by temperature and hydrogen partial pressure with the amount of Phase 1 and Phase 2 carbon increasing with increasing temperature and hydrogen partial pressure. The physical interpretation of this behavior is that as temperature increases, more bonds are broken, which means smaller fragments more amenable to gasification if they combine with hydrogen rather than recondensing. The increased hydrogen partial pressure maximizes the fraction of these thermally formed fragments that ultimately end up as gas.

As the above discussion indicates, coal is a chameleon-like substance that undergoes transformations in both chemical and physical properties that greatly affect its reactivity. Despite this complexity several models describing its reaction with hydrogen have been proposed that adequately describe the data.

For example, Johnson (*16*) assumed that the carbon conversion can be described by the following three rate regimes: (1) devolatilization, (2) rapid rate methane formation, and (3) low-rate gasification. Since devolatilization occurs nearly instantaneously, only the second two regimes can be described by rate equations. For the rapid rate methane formation period Johnson developed the following equation:

$$\int_0^{x_R} \frac{\exp (\alpha X^2) dX}{(1 - X)^{2/3}} = .0092 \, f_R P_{H_2}$$

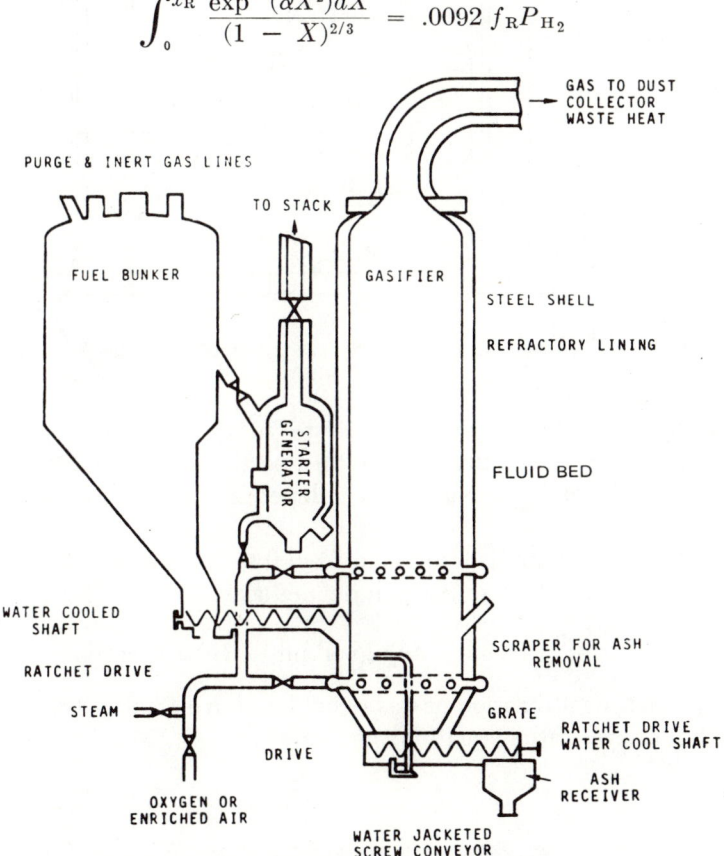

Figure 6. Winkler gasifier (courtesy Battelle Institute)

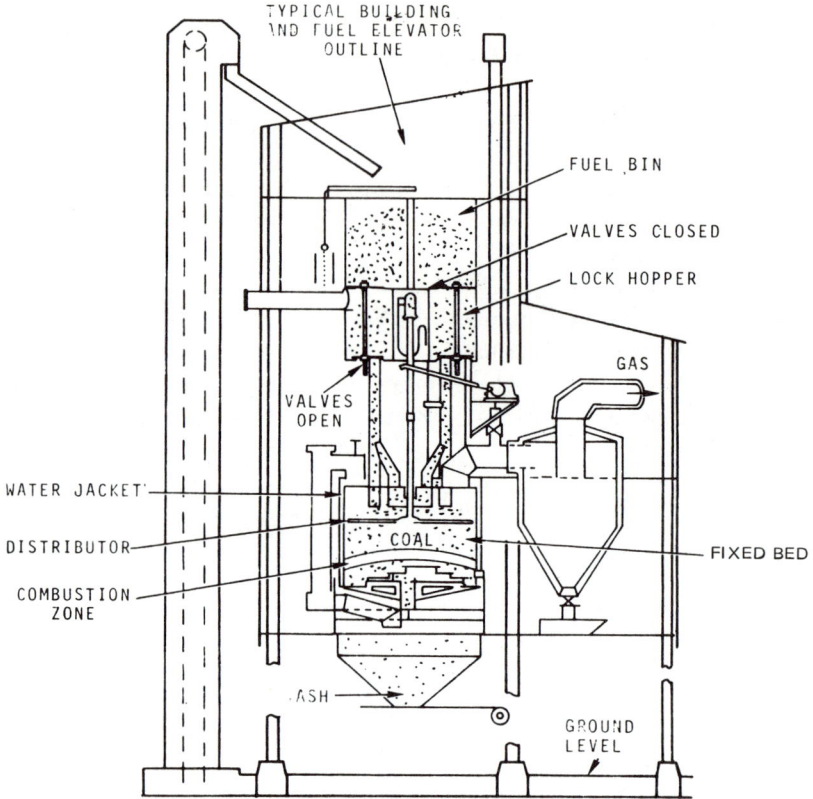

Figure 7. Wellman-Galusha fuel gas generator (courtesy Battelle Institute)

where P_{H_2} = hydrogen partial pressure in atmospheres; f_R = relative reactivity factor which depends on the particular type of coal or char used; α = a kinetic parameter that depends on gas composition and pressure.

Carbon conversion is integrated from 0 to X_R because the conversion is written in terms of "base" carbon defined as

$$\text{"base" carbon} = \text{total carbon} - \text{volatile carbon}$$

Johnson suggests utilization of this equation above 1500°F (816°C). The differential form of the equation is given by:

$$dX/dt = f_L k_T (1-X)^{2/3} \exp(-\alpha X^2)$$

where $(1-X)^{2/3}$ represents a surface area term and $(-\alpha X^2)$ represents the decay in reactivity experienced as the carbon conversion increases.

The rate constant, k_T, is in turn a complex function of reactant partial pressures and equilibrium constants (calculated on the basis of a carbon activity $= 1$) of the hydrogenation reaction expressed by stoichiometric Expression 3. Johnson uses this same basic equation, with appropriately modified constants, for the gasification of char with steam.

Zielke and Gorin (17, 18) recommend the following equations to correlate the reaction of char with steam–hydrogen mixtures:

$$R_{Total} = AP^n \qquad R_{CH_4} = DP^m$$

where R_{Total} is the total specific carbon conversion rate and R_{CH_4} is the rate of methane formation; P is the total system pressure and A, D, n, and m are empirical functions of carbon conversion and the steam/hydrogen ratio. The rate of carbon–steam reaction is calculated as $R_T - R_{CH_4}$. They also provide correlations showing apparent energies of activation (ranging from 40 to 75 kcal) with which to adjust the rate equations for temperature.

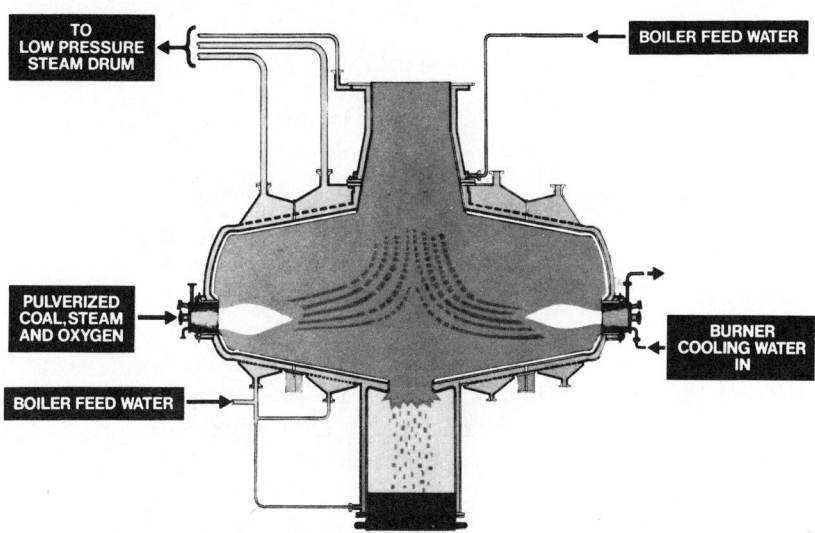

Figure 8. Koppers-Totzek gasifier

Zahradnik and Glenn (19) start the development of their hydrogasification rate equation with the following simplified reaction scheme:

$$\text{coal} + H_2 \rightarrow C^{**} + CH_4 + \text{(other gases)}$$

$$C^{**} \xrightarrow{R_1} C^*$$

$$C^{**} + H_2 \xrightarrow{R_1} CH_4$$

where C^{**} is an active intermediate; C^* is inactive char formed by polymerization of active intermediate forms of carbon. By assuming that the net conversion rate of active carbon is given by:

$$dC^{**}/dt = k_1 p_{H_2} A^{**} - k_2 C^{**}$$

where A^{**} is the area associated with the active intermediate and k_1 and k_2 are the rate constants for activation and deactivation. It was then assumed that the A^{**} term is proportional to $(C^{**})^{2/3}$. That is, the methane formation rate from the active carbon occurs at the shrinking surface surrounding the active species.

From this beginning the final equation relating methane yield to system parameters is

$$\text{methane yield} \frac{\text{lbs carbon in CH}_4}{\text{lbs C in original coal}} = b_1 + b_2 p_{H_2}/(1 + b_3 p_{H_2})$$

where b_1, b_2, and b_3 are constants determined by experimental measure-

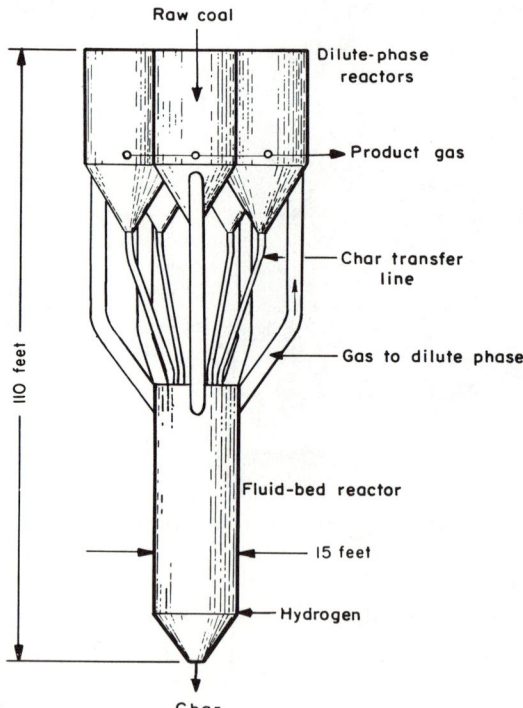

Figure 9. Commercial hydrane reactor. Capacity: 125 million SCF pipeline gas.

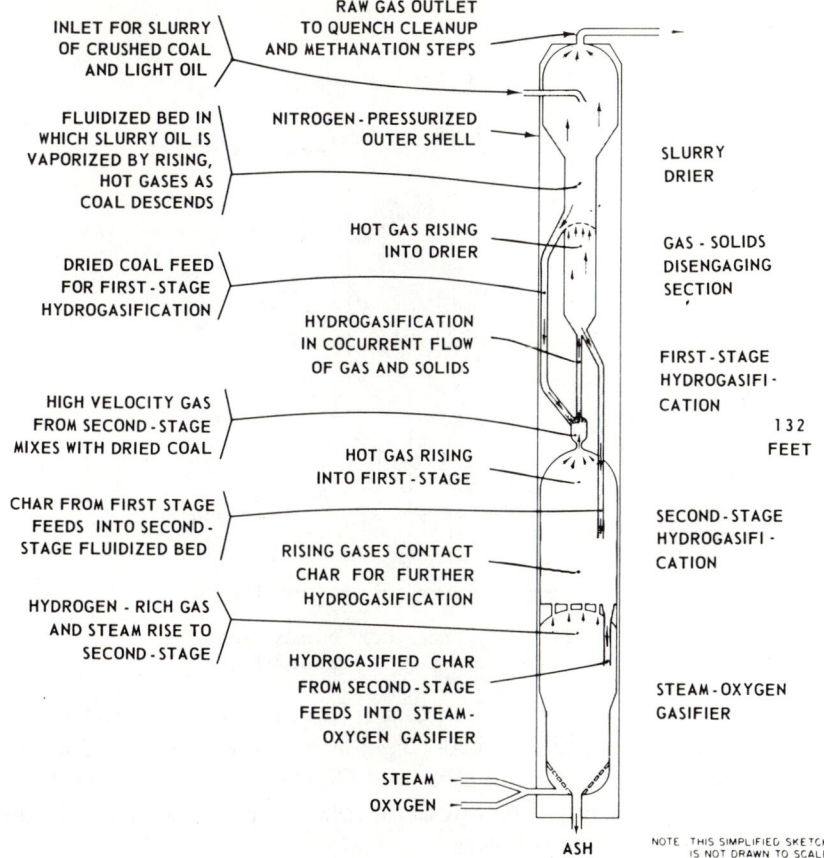

Figure 10. IGT pilot plant hydrogasification reactor section

ments. The constant b_2 is an Arrhenius function of temperature. This model was based on data generated in both integral and differential reactors.

Feldmann and co-workers (7, 14, 20) found a simple rate model adequate to correlate data from an integral dilute-phase reactor system using raw coal. The model assumes that the hydrogasification reaction proceeds throughout the particle and that a certain fraction of the carbon becomes converted into a form that is unreactive towards hydrogen. The equation which describes the rate of methane generation is:

$$dx/dt = kp_{H_2}(1 - x - Ci/Co)$$

where x = the carbon conversion to methane; k = the rate constant for methane formation; p_{H_2} = the hydrogen partial pressure; and Ci/Co =

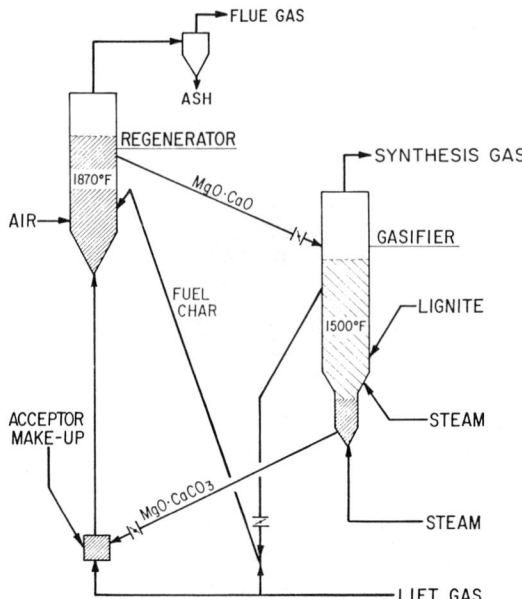

Figure 11. CO₂ acceptor process (courtesy American Gas Association)

the fraction of the original carbon that is deactivated. Over the range of carbon conversions achieved (20–50%) it appeared that the formation of unreactive carbon was negligible and $C_i/C_o = 0$. The constants were calculated from integral free-fall dilute phase reactor data by the integrating rate equation as follows:

$$\int_E^x dx/(1 - x - C_i/C_o) = k p_{H_2} L/U_T$$

where E = the fraction of carbon that is "instantaneously" converted to methane; L = the reactor length, and U_T = the average terminal velocity of the particles in the free-fall reactor.

Previous measurements (14) indicated that the gas is essentially backmixed in the reactor, thereby allowing a constant value to be assumed for the hydrogen partial pressure over the length of the reactor.

During these investigations k depended quite heavily on the type of reactor system used. For example, it was decided to use a differential bed reactor to generate lower temperature data than could be obtained in the pilot plant. Preliminary results from the differential reactor are compared with those measured in the integral free-fall dilute phase reactor in Figure 2. It is felt that the difference is caused either by the lower heatup rate in the differential reactor compared with the dilute phase reactor or

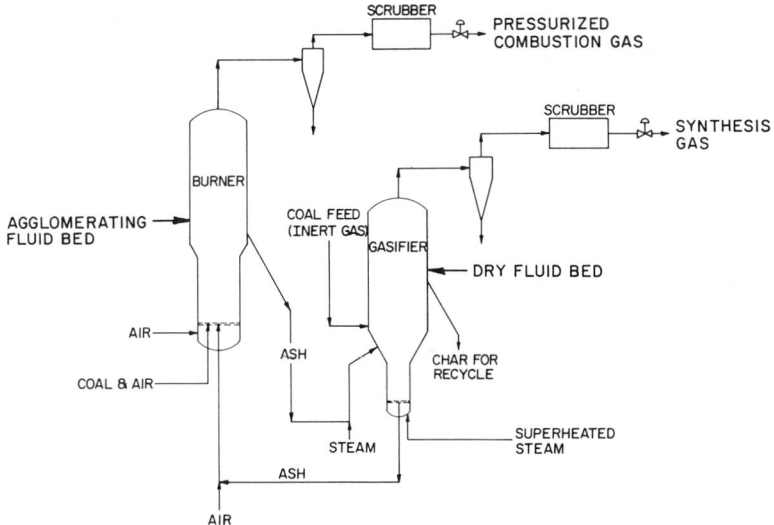

Figure 12. Union Carbide/Battelle agglomerating bed gasifier

the physical constraints on particle expansion in the differential reactor.

The porous nature of the char produced by the hydrogasification of raw coal is shown in Figure 3. This photomicrograph also indicates that the assumption of a shrinking core type model is probably not physically realistic. The softening and expansion of bituminous coal into porous spheres is shown in Figure 4. The agglomeration noted is caused by the particles colliding together while still in the sticky state.

Thus, we see that from a very similar physical concept various rate equations have been developed, all of which provide adequate fits to the experimental data at hand, and, with modification of appropriate constants, they can be extended to fit other data as well. Ordinarily, the number of experiments necessary to determine the constants for one of the models presented may be sufficient to develop an alternate rate model utilizable for process design.

Char–Steam Reaction ($C + H_2O \rightleftarrows CO + H_2$). This reaction is extremely important in determining the economics of gasification because it is endothermic, and (as previously discussed) the heat is usually supplied by combusting part of the carbon with oxygen. Also, the cost of oxygen constitutes one of the major costs of both synthesis gas and SNG. Many investigators have, over the years, studied the carbon–steam reaction and have proposed various mechanisms to explain their results. An excellent survey, discussion, and attempt to provide a unified picture of the system is given by Von Fredersdorff and Elliott (21).

Much of the work done to elucidate reaction mechanisms for the carbon–steam system utilized highly graphitic forms of carbon or pure chars such as that prepared from coconut shells. Unfortunately, it has not been possible to apply the resulting models to systems using coal and/or coal chars without a substantial number of experiments to define the constants used in the rate equations. Therefore, as in hydrogasification, there seems to be no safe way of avoiding direct experimental measurements utilizing the particular coal or char of interest. The resulting data may be correlated using a number of different reaction models. Some of these models are tabulated below.

Von Fredersdorff and Elliott (21) as well as many others recommend an equation of the Langmuir-Hinshelwood type

$$r_c = kp_{H_2O}/(1 + ap_{H_2} + bp_{H_2O})$$

to correlate carbon–steam reaction rate data. For example, this approach has been used successfully to correlate data on coal chars arising from lignite (22).

Wen and co-workers (23) in correlating data from a continuous flow reactor system consisting of coal char, steam, and hydrogen used a model bearing more physical resemblance to the models described previously for the coal–hydrogen system. That is, they assumed that the char consisted of two carbon phases, each having different reactivities toward steam. The more reactive first phase carbon in the char reacts at a rate given by

$$r_{c_1} = k\,(f - x)$$

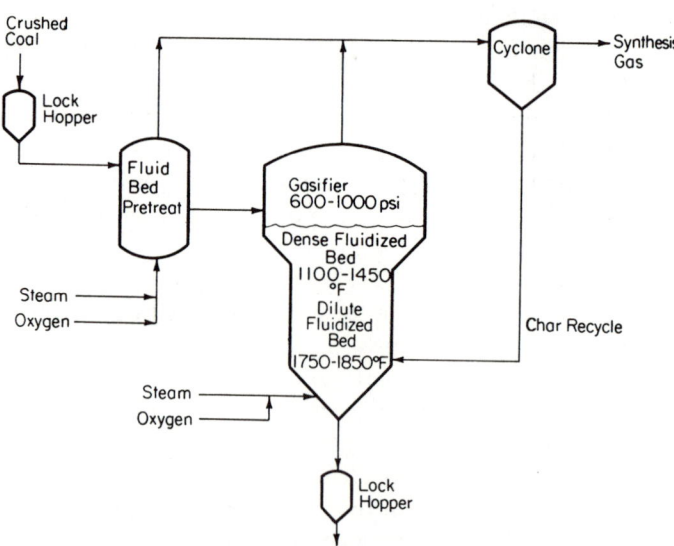

Figure 13. U.S. Bureau of Mines Synthane coal gasification process (courtesy Pittsburgh Energy Research Center—ERDA)

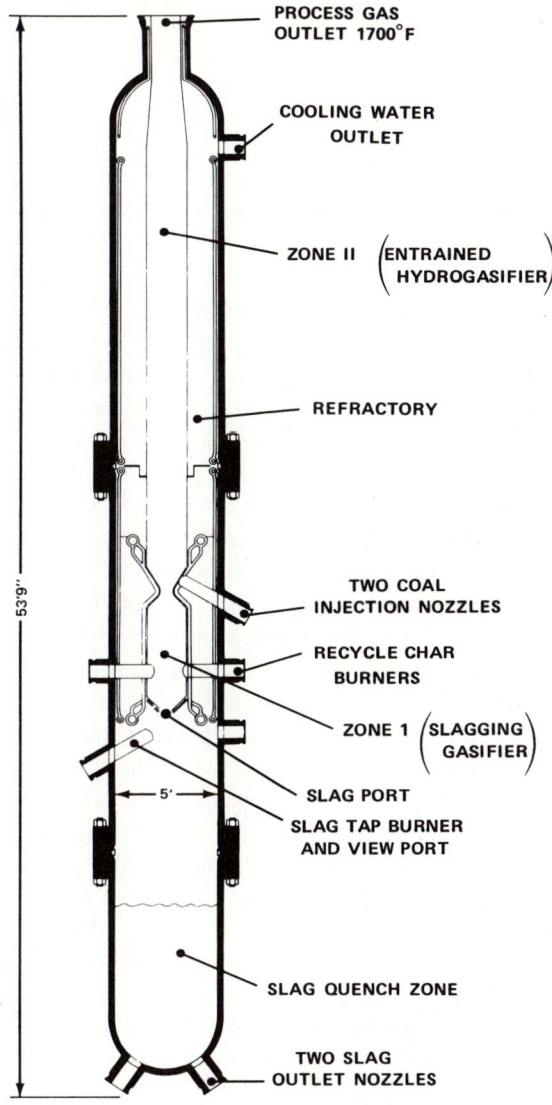

Figure 14. Bi-Gas reactor (courtesy American Gas Association)

where x is the carbon conversion and f is the fraction of first phase carbon in the coal or char feed. Thus, according to this model the gasification of this carbon is independent of steam partial pressure and is not retarded by hydrogen. Wen *et al.* (*23*) assume that the reaction of the second phase or less reactive carbon is diffusion controlled and recommend the rate expression

$$r_{c_2} = (1/D_p)[Kg_{H_2O}\,(1-f)\,(p_{H_2O} - p^*_{H_2O})]$$

where $(Kg)_{H_2O}$ is the effective mass transfer coefficient for steam, D_p the particle diameter and $p^*_{H_2O}$ the steam partial pressure in equilibrium with the gases surrounding the particle.

In Ref. 23 Wen and co-workers also present a model for coal hydrogasification that assumes two carbon phases, each having a different reactivity.

As mentioned, Johnson (*16*) uses the same basic rate equation to correlate the char–steam reaction as the carbon–hydrogen reaction with the constant, k, given by a different expression of temperature and gas properties. That all the above rate expressions allow the data to be adequately fitted is inevitable because of the number of constants at one's disposal. In gasification like hydrogasification the history of the char has an appreciable effect on its subsequent reactivity, and it is therefore necessary to conduct experiments to characterize the particular char of interest. In spite of the many efforts over the past years to develop gasification models, there is no *a priori* way of selecting constants to be used with any of the reaction models. Also, because of the complexity of the system, it seems that efforts to develop universally applicable kinetic models for design may be fruitless. Instead of developing additional kinetic models, further efforts ought to be increasing our qualitative understanding of what occurs during gasification and hydrogasification. For example, we have seen that in spite of the irreversibility of the reaction: coal + A(gas) → char + B(gas), many of the kinetic models successfully utilize approach to equilibrium arguments to correlate their data. We have also seen that, in spite of photomicrographs that indicate that char particles are extremely porous, shrinking core analogies have been utilized to develop rate equations that also allow the satisfactory correlation of data. Perhaps future hydrogasification rate studies should be placed in the following categories:

(a) Process design reaction rate models where the objective is to develop rate equations suitable for specific design purposes. To be suitable the data for such models should be generated at conditions simulating as closely as possible those that will be anticipated in the particular reactor system under design.

(b) Models that elucidate the physical phenomena occuring during hydrogasification. The emphasis in this work should be at increasing the understanding of what goes on rather than on the generation of mathematical rate models although the rates at which such transformations occur are important. One area needing such understanding is behavior of methane at gasification conditions. It is well known that the production of methane, the desired end product for SNG, is not favored at conditions optimal for gasification. Is this caused by (a) suppression of methane generation from coal or char at gasification conditions,

(b) steam reforming of methane, (c) carbon deposition catalyzed by char *via* the reaction: $CH_4 \rightarrow C + 2H_2$, or (d) combustion of methane with oxygen arising from high gas mixing rates?

Other areas of interest and importance would be

(a) steam decomposition by water–gas shift ($CO + H_2O \rightarrow CO_2 + H_2$) *vs.* steam decomposition by carbon–steam reaction ($C + H_2O \rightarrow CO + H_2$).

(b) more chemical and physical characterizations of char properties and their change with increasing carbon conversion.

(c) role of heatup rate in determining reactivity towards steam and hydrogen.

(d) reactivity of char carbon with steam at high carbon conversion levels. Many of the quantitative studies of the char–steam reaction were made of carbon conversion levels below 80%. Knowledge of the above factors should allow concepts for optimizing carbon and oxygen utilization—*i.e.*, maximum carbon conversion at minimum oxygen consumption to be developed.

(e) determination of intermediates in converting coal to methane.

(f) estimation of heat effects accompanying hydrogasification.

Knowledge of the above phenomena could lead to the design of more thermally efficient gasifiers.

Reactor Design. Almost every conceivable type of solids–gas contacting scheme is used in hydrogasification reactor systems, often in combination (Table I). For example, among high-Btu gas processes currently under development, the Hy-Gas Process uses a combination of fluid beds as well as an entrained reactor; the Bi-Gas Process utilizes an entrained reactor together with a slagging bed reactor system; the Battelle/Union Carbide Agglomerating Bed Gasifier utilizes a unique agglomerating type of fluid bed together with a two-stage non-agglomerating (dry) fluid bed; Consol's CO_2 Acceptor Process consists of two dry fluid beds very similar to the Battelle/Union Carbide reactor system; the Hydrane Process uses dilute solids phase free-fall reactors together with a fluid bed; and, finally, the Synthane Process requires a fluid bed and free-fall reaction zone. There are also four commercial systems that could be used to produce synthesis or high-Btu gas. Two of these proecsses—the Lurgi and the Wellman-Galusha gasifiers—use moving beds; the Winkler system uses a fluidized bed, and the Koppers-Totzek is an entrained reactor. All of these commercial reactors are single-staged, and only one, the Lurgi, is designed for operation at elevated pressure.

The primary reasons for the increased complexity of the reactor systems currently under development, compared with the aforementioned commercial systems, are to increase the fraction of methane produced by the hydrogasification, to reduce (or eliminate) oxygen consumption, and in the case of the Bi-Gas and Hydrane processes to allow the use of raw coal without pretreatment. Schematic representations of these reactor systems are shown in Figures 5 through 14. One of the major problems in operating the more sophisticated second generation hydrogasification systems now being developed will be the control of gas and solids flow at the high pressures and temperatures necessary for gasification and hydrogasification. Requirements to design such reactors and to control their operation will be:

(1) increased attention to the modeling of solids–gas flow circuits

(2) expanding the data and technology base for the design and control of such systems

(3) development of ways to predict gas mixing between reactor zones.

For example, the practical importance of the above topics in reactor design was recently demonstrated by Zahradnik and Grace (24) who showed that the injection of coal into the upper stage of the Bi-Gas reactor resulted in gas mixing patterns that caused methane to recirculate to the hot zone where it was reformed by steam. This greatly reduced the direct methane yield which has a deleterious effect on process economics. A simple change in the coal injection nozzle eliminated the problem and allowed increased methane yields. However, if gas mixing patterns in the reactor were not examined, this problem would not have been isolated, and the lower methane yields would have been accepted.

The specific references given in the Literature Cited section will provide additional information on coal gasification technology. The excellent comprehensive bibliography compiled by Ruby Mathison (25) is particularly recommended.

Literature Cited

1. Corder, W. C., *Proc. Synthetic Pipeline Gas Symp.*, *5th*, American Gas Association Catalog No. L51173, pp. 107-125, 1974.
2. Fink, Carl, *Ibid.*, pp. 89-107.
3. Tarman, Paul B., *Ibid.*, pp. 19-31.
4. Channabasappa, K. C., Linden, H. R., *Ind. Eng. Chem.* (1956) 48, 900-905.
5. Wen, C. Y., Li, C. T., Tscheng, S. H., O'Brien, W. S., "Comparison of Alternate Coal Gasification Processes for Pipeline Gas Production," Ann. AIChE Meetg., 65th, New York, Nov. 26-30, 1972.
6. Henry, J. P., Jr., Louks, B. M., "An Economic Comparison of Processes for Producing Pipeline Gas from Coal," *Am. Chem. Soc., Div. Fuel Chem., Preprints*, Chicago, Ill., Sept. 1970.
7. Feldmann, H. F., Wen, C. Y., Simons, W. H., Yavorsky, P. M., "Supplemental Pipeline Gas from Coal by the Hydrane Process," Natl. AIChE Meetg., 71st, Dallas, Tex., Feb. 20-23, 1973.
8. Rudolph, Paul F. H., *Proc. Synthetic Pipeline Gas Symp.*, *4th*, American Gas Association Catalog No. L11173, pp. 175-215, 1973.
9. Elgin, David C., *Proc. Synthetic Pipeline Gas Symp.*, *5th*, American Gas Association Catalog No. L51173, pp. 145-171, 1974.

10. Rudolph, Paul F. H., *Ibid.*, pp. 171-175.
11. Menster, M., O'Donnell, H. J., Ergun, S., Friedel, R. A., "Coal Gasification," L. G. Massey, Ed., ADVAN. CHEM. SER. (1974) **131**, 1-9.
12. Feldmann, H. F., Discussions on Coal Hydrogasification, Gordon Conference on Coal Science, New Hampton, N. H., July 1973.
13. Moseley, F., Patterson, D., *J. Inst. Fuel* (Sept. 1965) 378-391.
14. Feldmann, H. F., Simons, W. H., Mima, J. A., Hitshue, R. W., "Reaction Model for Bituminous Coal Hydrogasification in a Dilute Phase," *Am. Chem. Soc., Div. Fuel Chem., Preprints*, Chicago, Ill., Sept. 1970.
15. Hitshue, R. W., Friedman, S., Madden, R., *U.S. Bur. Mines, Rept. Invest.* **6376** (1964).
16. Johnson, J. L., "Coal Gasification," L. G. Massey, Ed., ADVAN. CHEM. SER. (1974) **131**, 145-178.
17. Zielke, C. W., Gorin, E., *Ind. Eng. Chem.* (1955) **47**, 820.
18. *Ibid.* (1957) **49**, 396.
19. Zahradnik, R. L., Glenn, R. A., "The Direct Methanation of Coal," *Amer. Chem. Soc., Div. Fuel Chem., Preprints*, New York, Sept. 1969.
20. Feldmann, H. F., Yavorsky, P. M., *Proc. Synthetic Pipeline Gas Symp., 5th*, American Gas Association Catalog No. **L51173**, pp. 287-310, 1974.
21. Von Fredersdorff, C. G., Elliott, M. A., "Lowry's Chemistry of Coal Utilization," Supplementary Volume, pp. 892-1022, Wiley, New York, 1963.
22. Curran, G. P., Fink, C. E., Gorin, E., *Ind. Eng. Chem., Process Design Develop.* (1969) **8** (4), 559-567.
23. Wen, C. Y., Abraham, O. C., Talwalkar, A. T., "Fuel Gasification," ADVAN. CHEM. SER. (1967) **69**, 168-185.
24. Zahradnik, R. L., Grace, R. J., "Coal Gasification," L. G. Massey, Ed., ADVAN. CHEM. SER. (1974) **131**, 126-144.
25. "Synthetic Fuel Research: A Bibliography," compiled by Ruby Mathison, American Gas Association Catalog No. **H01974**, Aug. 1973.

RECEIVED December 4, 1974.

7

Multiplicity, Stability, and Sensitivity of States in Chemically Reacting Systems— A Review

ROGER A. SCHMITZ

Department of Chemical Engineering, University of Illinois, Urbana, Champaign, Ill. 61801

> *Attention to the topics of multiplicity, stability and sensitivity of states in chemical reactors and catalyst particles has stemmed principally from publications by Van Heerden, Amundson, and Aris in the 1950's. These and subsequent studies through the 1960's were mainly theoretical in nature, dealing primarily with problems involving a single exothermic reaction. More recently, experimental research has borne out some of the theory, but most of the distributed models used in mathematical work have not been put to careful experimental tests. A parallel literature in other fields of application, including biology, combustion and electrochemistry, has developed, introducing a variety of problems and broadening considerably the scope of this subject area. This review surveys the literature on these topics with particular emphasis on the chemical engineering literature.*

The essential topics of this review are the multiplicity (or uniqueness) of steady states of open chemically reacting systems, the stability of states to small and large perturbations, and the sensitivity of them to parameter or input changes. The topics have their principal application in the design, startup, and control of the various types of continuous-flow chemical reactors encountered in the chemical and petroleum industries. Other areas of application, including combustion, biology, and electrochemistry, certainly are not new; in fact, some of them have earlier roots than do reactor applications in the published literature.

Common usage of the words multiplicity, stability, and sensitivity in the literature and throughout this paper is according to the following definitions. The multiplicity of steady states is the number of different sets of state variables at which the time rate of change of all state variables is identically zero for a fixed set of conditions or parameters. Gavalas (1) has shown that one should expect the multiplicity to be an odd number for reacting systems providing the chemical kinetic expressions satisfy some rather liberal restrictions. These states are described by the word steady only to signify that the time rate of change of state variables at such states is zero—not to signify stability. A steady state is stable if perturbations within an arbitrarily small neighborhood surrounding the state die away to zero. If even the smallest such perturbations grow, the steady state is unstable. In keeping with the usual vernacular, stability so defined should properly be termed local asymptotic stability, to distinguish it from other types such as global stability, which implies stability to an arbitrarily large disturbance. Sensitivity is somewhat less precisely defined. It describes a general situation in which small permanent changes in a parameter have a large effect on the steady state. Sensitivity may or may not be connected with steady-state multiplicity and instabilities.

The intent in this review is to portray as much as possible the many physicochemical situations and the variety of intriguing behavioral characteristics which have been described in the literature. The major thrust is toward chemical reactor applications, but a final section on other applications is included. Most of the papers on chemical reactors have focused either on the continuous-flow, well-stirred reactor (CSTR), on a single catalyst particle, or on tubular or fixed-bed reactors. Papers of a very general nature have been rare. Accordingly, separate sections are devoted to each of these three subjects. The fourth section contains a more general discussion of the effects of mixing and mathematical modeling. An attempt is made at appropriate places to put the papers to be presented in Session VII of the symposium volume (ADVANCES IN CHEMISTRY SERIES No. 133) in perspective.

Review papers from previous symposia, particularly the one by Ray (2), fill some of the gaps in the present review. In addition, books devoted to this subject by Gavalas (1) and Perlmutter (3) and a volume edited by Oppelt and Wicke (4) are good general references. Two additional books by Aris (5) and Denn (6), both of which emphasize these topics, will soon be available.

As an estimate, the number of literature references to papers on multiplicity and stability in chemical reactors contained in this review amounts to about 40% of those published in this area of application through the past two decades. The percentage cited in other areas of application is much smaller. Certainly, some important contributions, particularly those in the Russian literature, are missed in this review.

The CSTR

In addition to its practical role as an important and common type of industrial reactor, the CSTR, or more generally, the ideally mixed open reacting system, has been the cornerstone for this area of research. The necessary mathematical theorems and methods of analysis are standard; laboratory studies are relatively simple to conduct, and the results are easily interpreted. Consequently, our knowledge and understanding of CSTR behavior probably set the upper bound on our capabilities for exploring and understanding the behavior of other open reacting systems and suggest the questions one normally poses when investigating more complex distributed reaction models. [I am adhering to the usual convention and jargon according to which "distributed models" is taken to mean those which account for spatial variations of one or more of the dependent variables—as in most models of porous catalyst particles and tubular reactors. In "lumped models," such as the CSTR problem, no spatial dependence is considered.] In fact, it is tempting to conjecture that the qualitative features of the behavior of distributed systems may all be easily expected by analogy to CSTR behavior, the lure basically being the fact that the mathematical description for most distributed models reduces to that of a CSTR as dispersion parameters become large. Those findings that disprove this conjecture are of the greatest interest in studies of distributed systems and usually can be termed surprising. Therefore, any newcomer to this field is advised to acquaint himself fully with the status of knowledge regarding CSTR behavior before embarking on new problems.

The Classic Theoretical Problem of a Single Exothermic Reaction. The classic CSTR problem, introduced in papers by Van Heerden (*7, 8*), Bilous and Amundson (*9*), and Aris and Amundson (*10*) involves a single homogeneous exothermic reaction occurring in a well-stirred, continuously-fed reactor. [The paper by Van Heerden in 1953 was not actually the first to treat multiplicity and instabilities in chemical reactors, but no paper before it had made a significant impact by 1953. Publications by Liljenroth (*11*) and Wagner (*12*) are among those which preceded it.] The facts that (1) so simple a system can exhibit multiple steady states, unstable states, and sustained oscillatory outputs and that (2) the methods of Liapunov and Poincaré are well-suited for investigating the stability and transient characteristics of such processes were brought out in these early publications. These same characteristics were studied in numerous subsequent theoretical papers covering most of the conceivable variations of the classic problem and giving rise to a lengthy literature culminating in recent publications by Poore (*13*) and Uppal, Ray, and Poore (*14*).

For first-order Arrhenius kinetics, the material and energy balances for the classic problem take the dimensionless form of Equations 1 and 2.

$$\frac{dc_f}{d\theta} = 1 - c_f - Da\, c_f \exp\left[\gamma\left(1 - \frac{1}{t_f}\right)\right] \tag{1}$$

$$L\frac{dt_f}{d\theta} = 1 - t_f + \beta\, Da\, c_f \exp\left[\gamma\left(1 - \frac{1}{t_f}\right)\right] - \alpha(t_f - t_a) \tag{2}$$

[In the equations presented throughout this paper, I have incorporated first-order Arrhenius kinetics with a single chemical reaction for purposes of discussion and illustration. The vast majority of theoretical models have used this form; other kinetic descriptions require obvious and straightforward modifications.]

These equations are expressed in terms of six parameter groups L, Da, γ,

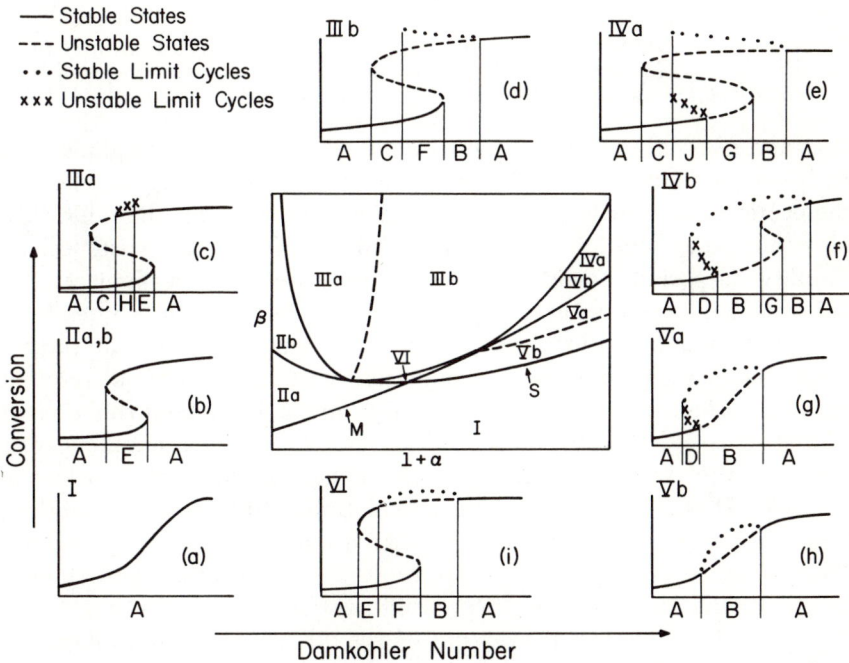

Figure 1. Steady-state and stability results for different regions in parameter space for a CSTR with first-order Arrhenius kinetics; L, γ, and t_a are fixed (*14*)

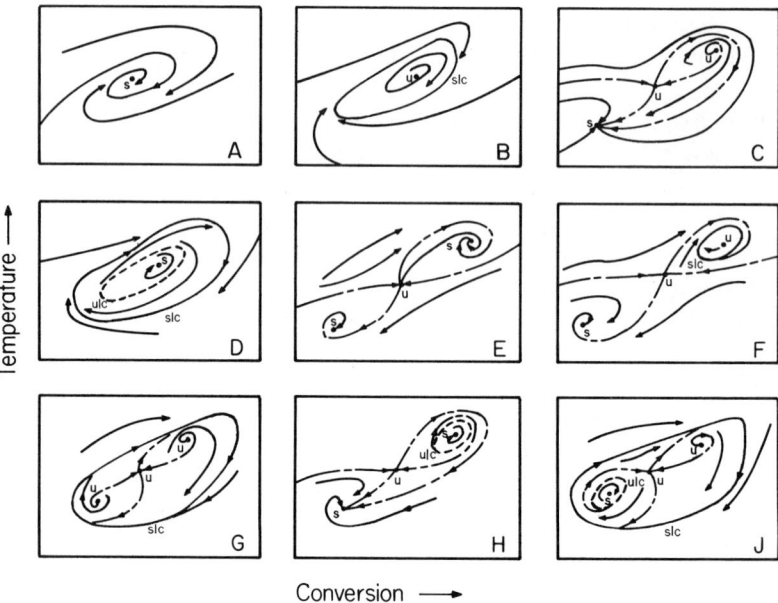

Figure 2. Classes of phase plots for the different cases indicated in Figure 1. The symbol u designates unstable steady states; s, stable steady states; slc, stable limit cycles; and ulc, unstable limit cycles (14).

β, α, and t_a. Poore (13) and Uppal et al. (14) established regions in the β, $1 + \alpha$ parameter plane (for fixed values of L, γ, and t_a) as shown in the center portion of Figure 1 by obtaining conditions for the multiplicity of steady-state solutions, for their stability and for the existence, and stability of periodic orbits (sustained osicillatory states—limit cycles). All of this information is summarized in Figure 1, which is intended to show only qualitative features. [In copying figures from the literature, I have changed notation and labeling from that in the original reference to be consistent with the discussion and symbols throughout this review.] The sketches of steady-state conversion vs. Damkohler number, which surround the central figure, typify the behavior within the various regions—I, II, IIIa, IIIb, etc. The curves bifurcating from the solid and dashed portions of the steady-state curves indicate the amplitudes of limit cycles; those marked by dots represent stable orbits and those by the symbol X, unstable ones. Nine different sections of the steady state curves, each distinguished from any of the others by the multiplicity and stability of steady states and the existence of one or more limit cycles, can be identified. These sections are designated A, B, . . . , H, and J on on the abcissas of the steady-state plots in Figure 1. For each of these sections there is a characteristic phase portrait in the conversion–temperature plane. Sketches of these are shown in Figure 2. If, for example, the parameters β and α (again for fixed values of other parameter groups) corresponded to a point in region IIIb of the central sketch in Figure 1, the behavioral features of that reactor are characterized by the sketch in Figure 1d. Phase portraits of type A, B, C, and F are possible, depending on the particular value of the Damkohler number. As the Damkohler number is decreased from large values, the branch of high conversion states is stable through section A and bifurcates to a limit cycle in section B. The limit cycle persists through section F, even though two other states—one a stable low-conversion state, the other a saddle point—emerge in the pase plane. Finally, the limit cycle disappears as it interferes with separatrices in the phase plane (at the right end of section C), and eventually only low conversion states at low values of Da are possible.

A sufficient condition for uniqueness of a steady state is that the point corresponding to given values of β and α lie below Curve M in the central sketch of Figure 1. All steady states for any point below Curve M satisfy the so-called slope condition for stability or static condition, as it was termed by Gilles and Hofmann (15). (This is the condition that the determinant of the coefficient matrix in linearized transient equations be positive or that the slope of the heat removal curve exceed that of the heat generation curve at the steady state.) For any point above Curve M, multiple states will exist over some range of Da. Stability of all steady states can be assured if and only if the point in the β, $1 + \alpha$ plane lies below both Curves M and S—i.e., in Region I. In fact, it can be shown that steady states for such cases are globally stable. All steady states for points below Curve S satisfy the condition that the trace of the coefficient matrix be negative—called the dynamic condition by Gilles and Hofmann (15). It can be shown that only Regions I and II are accessible in the special case of an adiabatic reactor (i.e., $\alpha = 0$).

It is not feasible to elaborate here on the abundant additional information contained in Figures 1 and 2. The paper by Uppal et al. contains a much more extensive discussion of reactor behavior for the different cases and of the methods of construction of Figure 1. It also includes numerical examples and simulations to add quantitative fidelity to the sketches shown here.

The work of Poore and Uppal et al. mainly serves to tie together all of the prior fragmentary information reported on this classic problem. Most of the separate features which they describe have been studied. The work of Van Heerden (7, 8) made it clear that either one or three steady states exist, depending on parameter values, and that the intermediate states are unstable. Subsequent studies (10, 15, 16) demonstrated that states other than the intermediate state could be unstable in nonadiabatic systems, and the phase diagrams of Aris and Amundson (10) showed that both stable and unstable limit cycles might be expected. Schmitz and Amundson (16) showed that all three steady states could be unstable, but the following facts established by Uppal et al. apparently have not been expressed in previous studies of CSTR behavior and were not generally appreciated by workers in this area. (1) For parameter values corresponding to points above Curve S in Figure 1, except for those in Region IIb, limit cycles will always exist over some range of values of the Damkohler number; (2) the stability of the limit cycles near the point of bifurcation can be determined by an algebraic criterion; (3) limit cycles may disappear as the Damkohler number is changed by (a) shrinking to zero amplitude (as, for example, in moving along the steady-state curve from Section B to Section A in Figures 1d, 1e, 1f, 1i, and 1h as stable limit cycles shrink to a stable state, or in moving from Section H to Section C in Figure 1c as unstable limit cycles shrink to an unstable steady state), (b) coalescence of stable and unstable limit cycles (as in moving from Region D to Region A in Figures 1f and 1g—generally referred to as "hard" bifurcations), and (c) interference of a limit cycle with a separatrix (as in moving from section H to section E in Figure 1c).

At least two other papers (17, 18) have presented an extensive analysis of this classic CSTR problem in parameter space. Neither is as exhaustive (particularly in the treatment of the appearance and disappearance of periodic orbits) as the publication by Uppal et al. Othmer (19) recently presented a similar analysis of a simplified kinetic model of the isothermal Belousov-Zhabotinskii reaction—an autocatalytic system in which malonic acid is oxidized isothermally by bromate in the presence of a metal ion. (Other works which focused on this fascinating reaction are cited later.) Also, the diagrams and analyses similar to those used by Uppal et al. can easily be constructed for other two-variable problems. With quite different kinetic expressions, new qualitative features could be introduced. For example, Uppal et al. mentioned the possibility of a greater number of limit cycles existing for a given set of parameters, but they found no evidence of these in their computations for first-order Arrhenius kinetics. Othmer (19) predicted the appearance of three limit cycles in a phase plane, two of which were unstable orbits around stable steady states and the third a stable orbit surrounding all three steady states. No simulations for such a case were given.

The exact construction of the phase diagrams shown in Figure 2 and the quantitative determination of the amplitudes of limit cycles (indicated schematically by the dots and X's in Figure 1) still require digital or analog computer solutions of the nonlinear Equations 1 and 2. Applications of the direct method of Liapunov to establish regions of asymptotic stability in the phase plane have consistently resulted in very conservative estimates, as well illustrated in a recent paper by Shastry and Fan (20).

Other references to such studies and a description of methods of constructing Liapunov functions may be found in the book by Perlmutter (3). Methods of obtaining the exact regions of asymptotic stability by computing that separatrix (by backward numerical integration) in the phase plane which passes through the saddle point have been described (10, 15) and are relatively easy to apply.

A number of papers have been devoted to methods of approximating the limit cycles in the phase plane (see, for example, Refs. 21, 22, 23). Heberling et al. (21) compared the various approximate solutions with those generated by numerical solution, including unstable limit cycles which were rendered stable by reversing the direction of time in the numerical integration. Much of the work on the computation of limit cycles has been motivated by the fact [first pointed out by Douglas and Rippin (24)] that the time-average performance given by an oscillatory output may be better than that of the steady state about which the output oscillates. Taking all experiences into account, most researchers in this field would probably recommend straightforward numerical or analog simulations using the nonlinear system equations to study the large scale transient effects for a given situation. [The periodic operation of chemical reactors caused by the deliberate manipulation (cycling) of one or more parameters to improve reactor performance is not within the scope of this review. A recent review by Bailey (25) covers this subject extensively.]

Further Theoretical Work. Among variations of or departures from the classic problem have been theoretical studies of systems involving more than one phase, more than one reaction, reactors in series, and others. Generally the features studied in these works are similar to those described above, but when the complexity of the system equations is increased (particularly as a result of an increase in the order of the system—*i.e.,* in the number of dependent variables), the multiplicity of steady states is often increased. Schmitz and Amundson (16) showed, for example, that a single exothermic reaction may give rise to as many as nine steady states when it occurs in both phases of a system with two fluid phases and with transfer resistances between the phases. In a similar situation involving multiphase polymerization, Goldstein and Amundson (26) encountered as many as 25 different steady states. (In connection with multiphase CSTR systems, there have been a few theoretical studies of steady-state multiplicity and stability in two-phase (27) and three-phase (28) models of fluidized-bed reactors with the dense phase perfectly mixed and of a perfectly mixed spray reactor (29), all for exothermic processes.)

A systematic analysis of two sequential reactions $A \rightarrow B \rightarrow C$, both exothermic, in a single-phase CSTR has been reported by Hlavacek, Kubicek, and Visnak (30); they showed that five steady states are possible, as are sustained oscillatory outputs for a nonadiabatic situation. The structure of the three-dimensional phase space for this system was calculated by Sabo and Dranoff (31), whose work also included the estimation of regions of asymptotic stability by application of Liapunov's direct method.

In a theoretical study which accompanied experimental work, Graziani et al. (32) analyzed the Belousov-Zhabotinskii reaction using a kinetic scheme very similar to the 10-reaction model of Field, Körös, and Noyes (33). The results, shown in Figure 3, predict "hard" oscillatory bifurcations (that is, the elimination of stable limit cycles by coalescence with unstable ones).

Many publications have focused on the problem of automatic control, specifically the possibility of stabilizing an inherently unstable state by a feedback control scheme. Some of the very early papers considered this problem (10, 34, 35, 36). More recently, stabilizing control has been considered by Demo et al. (37), who analyzed the possibility of stabilizing an unstable state by manipulating the stirring speed in a nonadiabatic reactor; by Hyun and Aris (38), who examined the effects of hysteresis in the feedback loop; and by Luyben (39), in a study of the effect of the reaction velocity constant on reactor stability.

Interest in the multiplicity and stability of steady states in isothermal

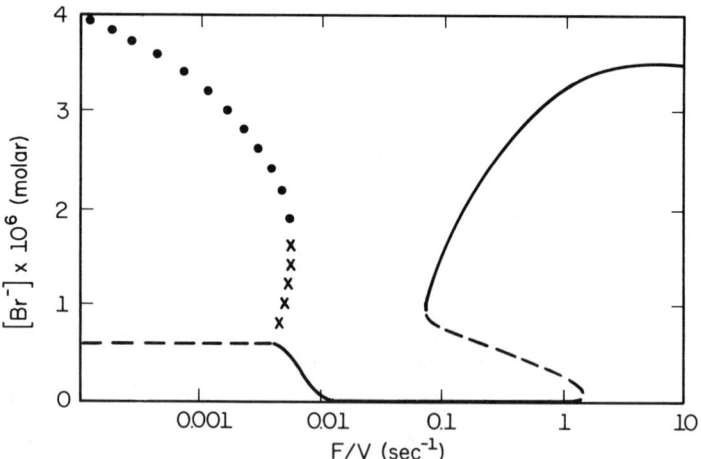

AIChE Meeting

Figure 3. Predicted behavior for the Belousov-Zhabotinskii reaction in a CSTR at 25°C. Solid curves represent stable states; dashed sections, unstable states. Dots represent the amplitudes of stable limit cycles (amplitudes not shown to scale) and X's, amplitudes of unstable ones (32).

systems has increased notably in recent years. Two studies of the Belousov-Zhabotinskii reaction have already been cited. In addition, Higgins (*40*) discussed various isothermal kinetic models indicating multiplicity of states and oscillatory behavior in many cases. His paper was aimed at biological reaction mechanisms, as were others (*see*, for example, Refs. *41, 42, 43*). Further discussion of multiplicity and stability in biology is included later. Matsuura and Kato (*44*) presented a theoretical study showing multiple states in an isothermal CSTR, basing their kinetic model on the autocatalytic oxidation of isopropyl alcohol. Knorr and O'Driscoll (*45*) pointed out that viscosity effects on the termination of the polymerization of styrene gives the appearance of autocatalycity and causes multiple steady states.

Future work in this area will probably be addressed more and more toward complicated systems of reactions. Hopefully this will lead to some means of categorizing or classifying reaction systems so that only the typical behavior of each class need be elucidated. Single reactions seem to be appropriately divided into two classes: (1) those for which chemical rates increase with an increase in the extent of reaction over some range of the extent (accelerating reactions) and (2) those for which rates monotonically decrease with increasing extent (decelerating reactions). To the former class belong exothermic and autocatalytic reactions as examples and to the latter, endothermic and isothermal mass-action reactions. Barring the possibility of some very unusual physical effects, a sufficient condition for the unique condition and stability of steady-state solutions is that the reaction belong to the decelerating class. Another important distinction between the two classes is that the progress of accelerating reactions may be favored by backmixing while decelerating reactions are best done under segregated or plug-flow conditions. There does not seem to be a useful analogy in terms of accelerating or decelerating systems of reactions, and the task of defining easily identifiable classifications appears formidable. Recent work by Horn (*46*), Horn and Jackson (*47*), and Feinberg and Horn (*48*) represents a step toward such a goal. Reaction mechanisms considered in those studies were isothermal systems which obey mass-action kinetics. The principal notions involved were those of weak reversibility and deficiency of a mechanism. The essential result of interest here is a theorem which states that a sufficient condition for a mass-action mechanism to have a unique and globally stable steady state in an open, perfectly mixed system is that it have a deficiency of zero and be weakly reversible. Both the deficiency and weak reversibility can be readily determined from the reaction mechanism and are independent of the reaction velocity constants. The proof involves an elegant mathematical structure essentially aimed at establishing conditions under which the nature of steady and transient states in an open reacting system is the same as that dictated by thermodynamic laws for closed systems near equilibrium. Fortunately for most interested readers the general results have been applied in a manner most easily appreciated in the last of the above three references as well as in three other papers by Horn (49). The results seem to be useful for models in which the system is open as a consequence of assuming that certain components are present in fictitiously constant concentration—an assumption commonly invoked in biological applications. However, this author's limited testing of the theorem with continuous flow reaction models suggests that it is quite conservative in its present form. Certainly variations and modifications of this approach as well as entirely different approaches will be forthcoming.

Experimental Studies. Though long delayed in time from the theoretical work of Van Heerden and others, a number of papers presenting experimental data on steady-state multiplicity and stability in the CSTR have appeared, mostly within the past three or four years. [For purposes of organization here, papers which include experimental data are referred to in separate subsections. Most of them also contain some contribution to or discussion of an underlying theory.] Table I gives a list of the above-mentioned papers along with a brief description of each system studied and the observations reported. [To my knowledge, the lists of experimental reactor studies given in tables in this review are complete. However, the bold claim of no omissions is not made and most likely would not be correct.] The list contains studies of single-phase systems (liquid and gas), of two-phase systems (emulsion polymerization, gas–liquid and gas–solid), of exothermic, and of isothermal processes. [The gas-solid cases represent solid-catalyzed reactions. These experiments used a recirculation reactor (or recycle or loop reactor) which approaches

Table I. Experimental Studies of Steady-State Multiplicity, Instabilities, and Control in Well-Mixed Reactors[a]

Reference	Experimental System	o	m	c
1. Hofmann, 1965 (50)	Decomposition of H_2O_2 in liquid phase in nonadiabatic CSTR	x		
2. Hafke and Gilles, 1968 (51)	Liquid phase oxidation of ethyl alcohol by H_2O_2 in nonadiabatic CSTR	x		
3. Hugo, 1968 (52)	Decomposition of N_2O on copper oxide catalyst in adiabatic circulating reactor	x		
4. Bush, 1969 (53)	Vapor phase chlorination of methyl chloride in nonadiabatic CSTR	x		
5. Furusawa et al., 1969 (54)	Hydration of propylene oxide in liquid phase in adiabatic CSTR		x	
6. Baccaro et al., 1970 (55)	Hydrolysis of acetyl chloride in liquid phase in nonadiabatic CSTR	x		

Table I. Continued

Reference	Experimental System	o	m	c
7. Hugo, 1970 (56)	Exothermic decomposition of N_2O on copper oxide catalyst and isothermal oxidation of CO on platinum catalyst in circulating reactor	x		
8. Vejtasa and Schmitz, 1970 (57)	Liquid phase reaction between $Na_2S_2O_3$ and H_2O_2 in adiabatic CSTR		x	
9. Gerrens et al., 1971 (58)	Emulsion polymerization of styrene in three-stage series of isothermal CSTR's (three stable states observed)		x	
10. Horak et al., 1971 (59)	Liquid phase autocatalytic reaction of bistrichlormethyl trisulfide with aniline in one- and two-stage isothermal CSTR system		x	
11. Hancock and Kenny, 1972 (60)	Chlorination of methyl alcohol in isothermal gas–liquid system	x		
12. Horak and Jiracek, 1972 (61)	Hydrogen–oxygen reaction on platinum catalyst in adiabatic recirculating reactor	x	x	
13. Hugo and Jakubith, 1972 (62)	Isothermal oxidation of carbon monoxide on platinum catalyst in recirculating reactor	x	x	
14. Lo and Cholette, 1972 (63)	Liquid phase reaction between $Na_2S_2O_3$ and H_2O_2 in series of adiabatic CSTR's		x	
15. Dubil and Gaube, 1973 (64)	Oxo reaction in a nonadiabatic reactor modeled as a CSTR	x		
16. Eckert, Hlavacek, and Marek, 1973 (65)	Oxidation of CO on CuO/Al_2O_3 catalyst in adiabatic circulating reactor	x	x	
17. Eckert et al., 1973 (66)	Oxidation of CO on CuO/Al_2O_3 catalyst in adiabatic circulating reactor	x	x	
18. Graziani et al., 1973 (32)	Liquid phase Belousov-Zhabotinskii reaction in isothermal CSTR	x		
19. Marek, 1973 (67)	Liquid phase Belousov-Zhabotinskii reaction in isothermal CSTR	x		
20. Chang and Schmitz, 1974 (68)	Liquid phase reaction between $Na_2S_2O_3$ and H_2O_2 in nonadiabatic CSTR	x	x	x
21. Ding, Sharma, and Luss, 1974 (69)	Chlorination of n-decane in gas–liquid mixture in nonadiabatic CSTR		x	
22. Guha and Agnew, 1974 (70)	Liquid phase reaction between $Na_2S_2O_3$ and H_2O_2 in adiabatic CSTR		x	
23. Weiss and John, 1974 (71)	Liquid phase formaldehyde condensation to carbohydrates is isothermal CSTR		x	

[a] Column headings: o stands for sustained oscillations, m for steady-state multiplicity, and c for feedback control of unstable states; x in the columns indicates the behavior observed.

CSTR behavior at high recirculation rates. For purposes of mathematical analysis, they are frequently described by the usual CSTR equations.] Both sustained oscillatory behavior and hysteresis effects resulting from multiple steady states have been studied, bearing out some of the theoretical results. Generally in the studies of steady-state multiplicity, two stable states were obtained. Exceptions were reported by Horak *et al.* (*59*), in which two CSTR's in series resulted in three stable states; by Horak and Jiracek (*61*), in which the high-conversion state as well as the intermediate state was unstable—a situation resembling type C of Figures 1 and 2; and in a study of the isothermal solid catalyzed oxidation of carbon monoxide by Hugo and Jakubith (*62*), in which sustained oscillations were observed around high conversion states, resembling situations of type F. In all other studies, only situations of types A, B, and E have been reported.

The data shown in Figure 4 are typical of the results of experiments by Graziani *et al.* (*32*) with the Belousov-Zhabotinskii reaction. The experimental conditions include the range over which both stable and unstable limit cycles were predicted according to the theoretical curve in

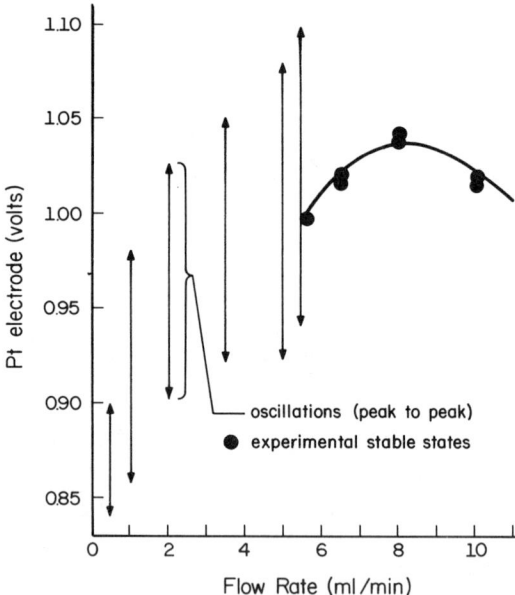

AIChE Meeting

Figure 4. Experimental results showing the readings from a platinum electrode immersed in a CSTR (volume 25 ml) for the Belousov-Zhabotinskii reaction at 25°C (32)

Figure 3. The data suggest that the oscillations are hard since they do not appear to shrink to zero amplitude at the end of the oscillatory region, and hence that situations of type D exist; however, extensive experimental testing did not establish this conclusively.

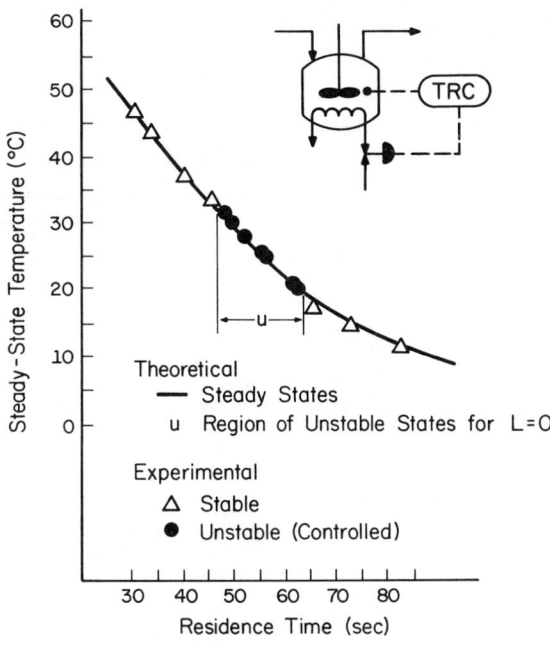

Figure 5. Experimental results of Chang and Schmitz (68) showing unique unstable (controlled) states for the thiosulfate–peroxide reaction in a nonadiabatic CSTR

Chang and Schmitz (68) used a simple feedback control scheme in a nonadiabatic system to stabilize unstable states—i.e., to eliminate the limit cycles in cases of sustained oscillations and to stabilize the intermediate states (saddle points) in the regions of multiple steady states—in an experimental analog of the early theoretical studies by Aris and Amundson (10). Figures 5, 6, and 7 illustrate some experimental observations of Chang and Schmitz and show some comparisons with theoretical predictions for the second-order reaction between sodium thiosulfate and hydrogen peroxide. Except for the automatic control aspect, these observations resemble those described in many of the other entries in Table I. Steady states on the range of residence times shown in Figure 5 are unique, but over a portion of that residence time range they are predicted to be unstable. The unstable states were obtained experimentally by a conventional proportional–integral feedback controller used as shown schematically in Figure 5.

Chang and Schmitz used two different reactors which differed only in their wall thickness—that is, in the effective value of L in Equation 2. For the data shown in Figure 5, L was estimated to be about 1.02. For the other reactor L was *ca.* 1.1—large enough to cause all of the steady states in Figure 5 to be stable without feedback control.

Figure 6 shows an experimental limit cycle for the uncontrolled reactor on the phase plane and the course of the transient following the initiation of feedback control action. [The phase plane shows the temperature *vs.* its time derivative as opposed to the usual temperature–concentration plots. Transient concentrations were not measured in these experiments.] Figure 7 presents theoretical and experimental results for

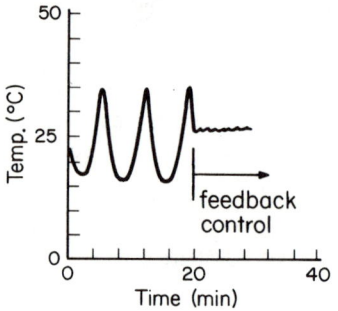

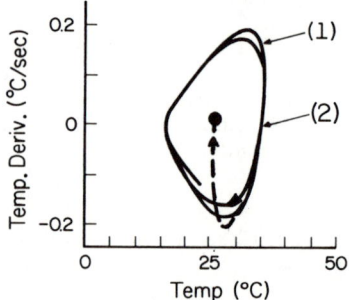

Chemical Engineering Science

Figure 6. Sustained oscillations and the effect of feedback control with the thiosulfate–peroxide reaction in a CSTR. In the lower diagram, Curve (1) represents the limit cycle, and the dashed curve starting at (2) shows the trajectory following the initiation of feedback control (68).

conditions under which multiple steady states exist. The experimental points for the intermediate states were again obtained by stabilizing feedback control. Both the high and low temperature states in Figure 5 were inherently stable.

In those cases for which comparisons were made, agreement between theoretical predictions and experimental results have been generally satisfactory, at least in a qualitative sense. An accurate description of the chemical kinetics is undoubtedly the limiting factor in this regard in almost any situation. A good example is the work of Hugo and Jakubith (62) on the oscillatory behavior in the solid-catalyzed oxidation of carbon monoxide. Other papers, though not reporting expeimental data, have closely linked their analysis to actual reaction systems or at least used real kinetic data (36, 44, 45, 72, 73, 74, 75, 76, 77, 78). Limited space does not permit a description or listing of the various topics covered in these papers.

The Catalyst Particle Problem

Mathematical Models and Theoretical Studies. The starting point for most theoretical studies of chemical reaction with thermal effects in porous catalyst particles has been the following system of dimensionless diffusion equations, written here for first-order Arrhenius kinetics:

$$\frac{\partial c_s}{\partial \theta'} = \left(\frac{\partial^2 c_s}{\partial x^2} + \frac{a}{x}\frac{\partial c_s}{\partial x}\right) - \phi^2 c_s \exp\left[\gamma\left(1 - \frac{1}{t_s}\right)\right] \quad (3)$$

$$Le \frac{\partial t_s}{\partial \theta'} = \left(\frac{\partial^2 t_s}{\partial x^2} + \frac{a}{x}\frac{\partial t_s}{\partial x}\right) + \beta'\phi^2 c_s \exp\left[\gamma\left(1 - \frac{1}{t_s}\right)\right] \quad (4)$$

where

$$a = \begin{cases} 0 & \text{for slab geometry} \\ 1 & \text{cylindrical} \\ 2 & \text{spherical} \end{cases}$$

with boundary conditions:

$$x = 0: \frac{\partial c_s}{\partial x} = \frac{\partial t_s}{\partial x} = 0 \text{ (for symmetric profiles)} \quad (5)$$

$$\left. \begin{aligned} x = 1: c_s &= c_f - \frac{1}{Sh}\frac{\partial c_s}{\partial x} \\ t_s &= t_f - \frac{1}{Nu}\frac{\partial t_s}{\partial x} \end{aligned} \right\} \quad (6)$$

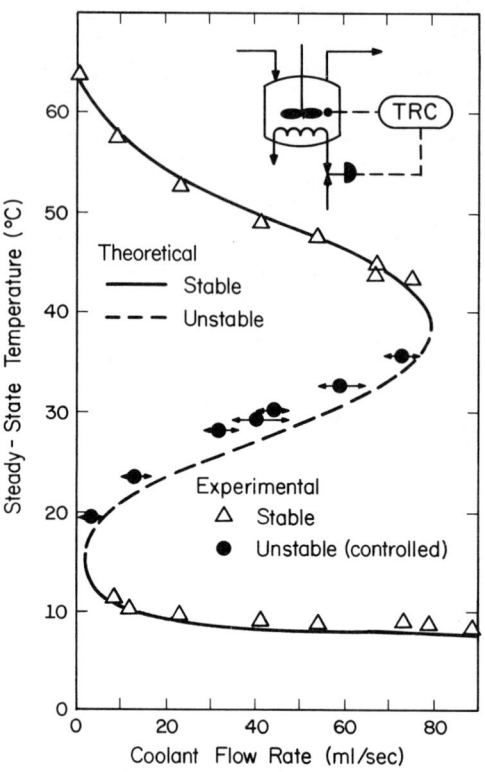

Figure 7. Stable and unstable (controlled) states for the thiosulfate–peroxide reaction in a CSTR. The arrowheads extending from the unstable states indicate the extent of fluctuations in the reactor temperature during controlled operation (68).

The quantity ultimately of interest in most applications is the particle effectiveness factor, η, which is obtained from the solution profiles and depends, of course, on all parameters in the describing equations.

Few studies have used these equations as written. Various special cases are deducible depending, for example, on whether (1) the Lewis number, Le, is taken to be unity, (2) slab, cylindrical or spherical geometry is adopted, (3) the Nusselt and Sherwood numbers are finite, (4) the Nusselt and Sherwood numbers are equal, (5) internal concentration and/or temperature gradients are considered and (6) the assumption of symmetric profiles, embodied in Equation 5, is retained. Most of the combinations of these special cases either make sense physically in certain applications or lead to mathematically convenient problems. Obviously the number of different versions of the basic problem is large; a literature search would show that questions regarding steady-state multiplicity and stability have been pursued for many of these. Some versions lend themselves to special mathematical methods not applicable to others, and certain distinctive properties of solutions of some are not found in others. All in all, it is difficult in a brief review to organize and sort out the applicable mathematical methods, as well as the effects of the many parameters, and to include conclusive statements, many of which must be qualified by a list of restrictions. The following account is intended only to bring out essential facts and, where possible, to relate behavioral characteristics for this problem to those of the CSTR. A very comprehensive treatment of the porous catalyst particle problem may be found in the book by Aris (5).

For typical commercial porous catalysts used in industrial reactors some of the special cases indicated above would not be realistic. For example, the ratio Sh/Nu can cover several orders of magnitude with the lower limit being approximately 10. Le is also much greater than unity—an average of perhaps 30. (Parameters values for some specific exothermic reactions were tabulated by Hlavacek et al. (80).) Such considerations lead to a practical working model in which the temperature is assumed to be uniform throughout the particle, any heat exchange resistance being confined to an outer fluid film, and in which all mass transfer resistance is internal. Because of the large Le, it is reasonable in the working model also to assume that the concentration profile is in a pseudo-steady state at all times—that is, that it is governed by the steady form of Equation 3. More is said later regarding this model. Most theoretical studies have not been restricted to these practical ranges of conditions and justifiably so, for one should adopt a broad view of the problem as essentially being a mathematical description of any two state variables in a general reaction–diffusion problem. For example, with appropriate kinetic expressions and interpretation of the various parameter groups, the equations shown above may describe the concentration profiles of two species in an isothermal system for which the magnitudes of parameter groups corresponding to Le and Sh/Nu may be greater than, less than, or equal to unity.

The catalyst particle problem described by Equations 3–6 bears a great deal of similarity to the CSTR problem, the flow terms in the latter replaced by diffusion terms in the former. Aris (81) explored these similarities in studying the problem for $Nu, Sh \to \infty$. Some parameters such as Nu, Sh and a, do not have obvious counterparts in the CSTR problem, but one would expect a priori that, like the CSTR problem, the catalyst particle would possess multiple steady states and unstable states under some conditions. If, in fact, internal gradients are neglected entirely or if reaction is assumed to occur only on the external surface and the temperature of the solid is assumed uniform, the mathematical problem takes on a form identical to that of the CSTR. The multiplicity of states and stability considerations for this version of the problem were studied by Wagner (12) and others (see, for example, Refs. 82, 83, and 86). Computations of steady-state profiles by Weisz and Hicks (87), Ostergard (88), Amundson and Raymond (89), and others, from equations similar to Equations 3 and 4, have demonstrated the multiplicity of states and hence the multiplicity of the effectiveness factor when intraparticle effects are involved.

Some studies have been aimed at establishing sufficient conditions for uniqueness and in some instances sufficient conditions for stability as well. Such conditions are potentially valuable because the computations involved in searching for all possible steady profiles require the solution of a nonlinear boundary value problem—a frequently difficult and expensive task. Ascertaining the stability of a given profile is even more problematic. In a practical situation, if one can rule out the possibility of multiple solutions and instabilities by easily applied sufficient conditions, the design and operation of a catalytic reactor is somewhat simplified. Many such conditions are available—all restricted to certain cases and some untested insofar as their conservatism is concerned. It is only a slight exaggeration to claim that the listing and comparing of these and mentioning the various mathematical methods put to use in deriving them would require a review paper itself. To cite some of these: criteria given by Luss (90) for $Nu, Sh \to \infty$ appear to be the least conservative and the most general in the sense that they apply to the various geometries and that an arbitrary form of the rate expression can be accommodated; Jackson (91) presented sufficient conditions for uniqueness for a case of finite Nu and Sh for slab geometry. Kastenberg (92) showed that Luss' critrion for uniqueness with $Nu, Sh \to \infty$ also guaranteed global stability for $Le = 1$. Liou et al. (93) obtained sufficient conditions for stability for $Le \neq 1$.

Aware of the results of the CSTR problem, one would expect the multiplicity of solutions of the steady-state forms of Equations 3–6 to be three over some ranges of the parameters and unity over others. While this is the case with $Nu, Sh \to \infty$ for the infinite slab and infinite cylinder geometries, it is not necessarily the case for spheres. Taking $Nu, Sh \to \infty$ for a spherical particle, Copelowitz and Aris (95) found as many as 15 steady solution profiles, and Michelsen and Villadsen (96) found 21 profiles. This great multiplicity was found only over a very narrow parameter range in both cases. Earlier, Hlavacek and Marek (97) had found five steady profiles for spherical geometry with a zero-order kinetic model.

Michelsen and Villadsen (96) showed that there may be infinitely many states in the spherical particle, a conclusion reached also by Gel'fand (98) and Fujita (99) in studies of a closely related mathematical problem. All of the solutions obtained by Copelowitz and Aris and by Michelsen and Villadsen have been shown subsequently to be unstable except for the high and low conversion profiles (96, 100). Even though this great multiplicity exists only over a very narrow and impractical range of parameters and all but two profiles are unstable, the fact that they do occur is of great interest because (1) neither geometry effects nor multiplicities greater than three for a problem of this type are suggested by CSTR results and (2) they alert researchers to the fact that while the well-studied CSTR problem provides tremendous theoretical insight, not all qualitative features of the behavior of open chemically reacting systems in general are contained therein.

Other theoretical discoveries, labelled as surprising when viewed against CSTR behavior, are: (1) that five steady-state solutions exist under some conditions when $Nu/Sh < 1$ (101) and (2) that if (a) the imposed condition in Equation 5 at the particle center is removed and the condition given in Equation 6 is imposed instead at $x = -1$ for slab geometry, (b) $Sh > Nu$, and both are finite and (c) the problem as described by Equations 3–6 has three steady-state solutions, then the slab may have six additional profiles (actually three sets of mirror-image pairs) which are not symmetric about the centerline. The possible existence of such profiles was shown by Pis'men and Kharats (102) and Horn et al. (103). The asymmetric profiles have also been the subject of other recent papers (104, 105, 106, 107, 108, 109). This result is apparently geometry dependent; no analogous findings for the infinite cylinder or the sphere have been reported.

A rigorous study of the stability of steady-state solutions is unwieldly, calling for an analysis of the eigenvalues of a nonself-adjoint system in which coefficients are functions of steady-state profiles. The problem is simplified in cases for which Le is taken to be infinity, unity or zero with $Nu = Sh$ because all eigenvalues can be shown to be real, and well-known mathematical theorems can be applied. Consideration of large-scale transients requires studying the solutions of coupled nonlinear partial differential equations in general. Some examples of stability analyses and transient simulations involving various analytical and numerical tools and based on a number of variations of the problem expressed by Equations 3–6 are contained in Refs. 83, 84, 85, 89, 100, 110, 111, 112, 113, 114). Included in some of these (112, 113) are numerical solutions for cases with $Le < 1$ having a unique unstable profile about which the state variables oscillate continuously. The oscillations resemble limit cycles in the CSTR problem. In general, studies of steady-state stability and of large scale transient behavior for the catalyst particle problem have shown no behavior that is not analogous to CSTR results. This *posteriori* observation is not meaningless because (1) there is no general theorem for distributed models which ensures that linearized equations yield the correct answers even for infinitesimal perturbations as do Liapunov's theorems for lumped models and (2) Poincaré's theorems guaranteeing limit cycles in a phase plane for second-order lumped models having a unique unstable state do not necessarily carry over to distributed problems.

Certain physical effects not described in Equations 3–6 have been studied in some publications. These include nonuniform external conditions (*94, 115*) to which a particle in a fixed bed reactor will generally be subjected and changes in molar density as a result of chemical reaction (*116, 117*). These and other such effects would be expected to give rise only to quantitative changes, which may be important in practical situations but would not be expected to (and have not been shown to) add or take away any characteristic features of the properties of solutions to Equations 3–6. The paper by Jackson (*117*) is of fundamental interest, however, because it points out that additional terms may belong in the transient equations if a change in molarity is associated with the chemical reaction.

It was pointed out above that even though wide parameter ranges have been explored in theoretical studies, practical considerations limit these ranges considerably for nonisothermal problems. A very useful, simple and plausible model for practical purposes may be deduced from Equations 3–6 if it is assumed that all of the resistance to heat exchange resides in the external fluid film and that the species profile is always in a pseudo-steady state relative to the instantaneous temperature profile. Invoking these assumptions, one may integrate Equation 4 term by term over the particle volume to obtain

$$Le \frac{dt_s}{d\theta'} = (a+1)\, Nu\, (t_f - t_s) + \beta' \phi^2 \eta\, c_f \exp\left[\gamma\left(1 - \frac{1}{t_f}\right)\right] \quad (7)$$

where η, the effectiveness factor, is given by any of the following three expressions,

$$\eta\, c_f \exp\left[\gamma\left(1 - \frac{1}{t_f}\right)\right] = \begin{cases} (a+1)\exp\left[\gamma\left(1 - \frac{1}{t_s}\right)\right]\int_0^1 c_s(x) x^a dx & (8a) \\ [(a+1)/\phi^2](\partial c_s/\partial x)_{x=1} & (8b) \\ [(a+1)/\phi^2]\, Sh\, [c_f - c_s(1)] & (8c) \end{cases}$$

where $c(x)$, $(\partial c/\partial x)_{x=1}$ or $c(1)$ for Equations 8a, 8b, and 8c may be obtained analytically for a given value of t_s from

$$\frac{\partial^2 c_s}{\partial x^2} + \frac{a}{x}\frac{\partial c_s}{\partial x} = \phi^2 c_s \exp\left[\gamma\left(1 - \frac{1}{t_s}\right)\right] \qquad (9)$$

In practical applications, η may be available empirically in terms of the parameters. These equations form a simplified particle model which might well be the most appropriate working model in commercial applications. Similar equations could be developed for kinetics other than the first-order Arrhenius type used here. The major points are: (1) Equation 7 is analogous to the energy equation for a CSTR, (2) steady-state values of η and t_s may be determined graphically if η is given in terms of t_s, and thus the possibility of multiplicity of steady states can easily be examined, and (3) a transient analysis using Equation 7 is relatively simple and shows that (a) the slope condition gives a necessary and sufficient test for stability, (b) no sustained oscillatory states are possible, and (c) unique steady states are globally stable. Numerical computations and comparisons by Hansen (118) and Yang (114) indicate the adequacy and limitations of this model. In particular, they show that while temperature uniformity in the steady state may be a good approximation, the transient profiles can show considerable spatial variation which could affect stability. This same model as well as other forms based on the notion of temperature uniformity through the particle have also been used in studies of steady state multiplicity and stability by Cresswell (119), Cresswell and Paterson (120), and McGreavy and co-workers (121, 122, 123, 124, 125, 126). McGreavy and Adderley (127) used this model to examine the steady-state sensitivity of the catalyst pellet to external parameters and related the sensitivity to the runaway problem in a fixed-bed catalytic reactor. More will be said regarding the runaway problem later.

An appreciable amount of work has been reported for the porous catalyst particle with models of chemical kinetics other than a single exothermic reaction. Gavalas (1, 111) showed that in general the multiplicity of steady-state solutions should be an odd number—say $2m + 1$, and that at least m of them will be unstable. Kuo and Amundson (83, 84) studied the multiplicity and stability of the sequential system of exothermic reactions $A \rightarrow B \rightarrow C$ and found five steady-state profiles in some cases. Cresswell and Paterson (120) and McGreavy and Thornton (121, 122, 126) also worked with models which accounted for more than one reaction in studying multiplicity, stability, and sensitivity problems. Hartman et al. (128) incorporated a dual-site isothermal Langmuir-Hinshelwood model into the diffusion equations and showed that three steady states were possible. Similar problems were worked by Mitshka and Schneider (129, 130). Such reactions, not obviously autocatalytic in a strict sense, show accelerating behavior as the reaction extent increases owing to a depletion of a reactant which tends to dominate the active sites.

Mehta and Aris (*131, 132*) presented a detailed study of the pth-order isothermal reaction and showed multiplicity for $p < 0$.

Lumped Models. The emphasis through the preceding section was on theoretical studies which used Equations 3–6 (or of special cases deduced therefrom) and which used methods of analyzing them or of generating numerical solutions. In recent years much effort has been devoted to the development and application of methods which convert the basic distributed model to an approximate lumped model—a process to which the term lumping has been applied. [Some workers have used the term lumping to indicate the replacement of a partial differential equation in an approximate sense by a single ordinary differential equation. It seems more consistent with usual terminology to refer to lumping generally as the process of converting a distributed system to a lumped system, even though the number of equations in the latter may exceed that of the former.] Essentially these methods involve replacing those terms containing derivatives with respect to position by a linear combination of the dependent variable evaluated at specific spatial positions. The number N of such positions is arbitrary except insofar as limitations may be placed on the accuracy of solutions; presumably as N is increased, the exact solution is approached. Thus a single partial differential equation is converted to N ordinary differential equations. The advantages of working with the lumped version are obvious: the knowledge gained from studies of the CSTR problem carry over directly, and mathematical tools for studying new behavior are readily available.

Four methods of lumping have been described: (1) linearization, (2) difference methods, (3) orthogonal collocation, and (4) averaging. All four have been applied to the catalyst particle problem in connection with studies of steady-state multiplicity and stability. The orthogonal collocation method has received the greatest amount of attention. It was first used for studying the catalyst particle problem by Villadsen and Stewart (*133*), who showed that spatial derivative operators could be replaced by matrices, the elements of which depend on the collocation points. The collocation points are usually taken to be the roots of an orthogonal polynomial. Orthogonal collocation is one of the methods of weighted residuals and is described in detail, with examples, by Finlayson (*134*). The first application of the method to steady-state multiplicity in a catalyst particle was reported by Stewart and Villadsen (*135*). It has subsequently been used in many studies of the catalyst particle problem, including those reported in Refs. *136, 137, 138, 139, 140, 141, 142, 143*, which addressed problems of steady-state multiplicity, stability, and total transient behavior. In most cases tested, the original partial differential equations may be represented by a rather low order system of ordinary differential equations by the collocation method. Rarely are more than three or four collocation points required, although Van Den Bosch and Padmanabhan (*144*) found it necessary to use a 14-point representation to attain good accuracy in calculating very steep steady profiles and a seventh-order system for stability determinations.

Linearization and difference methods have been used mainly by Hlavacek and others at the Prague school (*79, 145, 146, 147, 148*). [Linearization here is the term associated with the particular lumping method proposed by Hlavacek and co-workers in the papers cited. It has nothing to do with linearizing nonlinear terms in the system equations.]

The averaging method was introduced by Luss and Lee (*149*) in examining the stability of the steady-state profiles for a particle whose temperature was assumed uniform. It basically involves replacing the terms in the original distributed model by quantities averaged over the particle volume or over subsections of it. The spatial derivative operators take on the form of overall transfer coefficients. This method was also used by Lee et al. (*150*).

In many of the publications cited above, comparisons are made of one lumping method vs. others or of results and computational labor associated with lumping methods vs. those of other methods of handling the original partial differential equations (see, for example, Refs. *79, 80, 140, 141, 142, 143, 144, 146, 147, 149*). A very interesting and promising observation is that first-order lumping is suitable for many purposes. In first-order lumping with the collocation method, for example, one would use only one interior collocation point. In this approach the partial differential equations in Equations 3 and 4 are each replaced by a single ordinary differential equation, the spatial derivative term in Equation 3 replaced by $\lambda_c(1 - c_s)$, and in Equation 4 by $\lambda_t(1 - t_s)$. The coefficients λ_c and λ_t depend on the particular lumping method used, and c_s and t_s are considered to be average values or values at a specific point. In the orthogonal collocation method, c_s and t_s are approximations to the exact solutions at the single interior collocation point, and λ_c and λ_t are given by the following relationships

$$\lambda_c = n_1 \left(\frac{Sh}{Sh + n_2}\right); \lambda_t = n_1 \left(\frac{Nu}{Nu + n_2}\right) \tag{10}$$

where n_1 and n_2 depend on the particle geometry. Equations 3–6 then reduce simply to

$$\frac{dc_s}{d\theta'} = \lambda_c (1 - c) - \phi^2 c_s \exp\left[\gamma \left(1 - \frac{1}{t_s}\right)\right] \tag{11}$$

and

$$Le \frac{dt_s}{d\theta'} = \lambda_t (1 - t_s) + \beta' \phi^2 c_s \exp\left[\gamma \left(1 - \frac{1}{t_s}\right)\right] \quad (12)$$

These are immediately recognized as being equivalent to the adiabatic CSTR problem; in fact, they can readily be put in the form of Equations 1 and 2 with L replaced by $Le\, \lambda_c/\lambda_t$, Da by ϕ^2/λ_c, and β by $\beta'\lambda_c/\lambda_t$. Then all of the information contained in Figures 1 and 2 and in the earlier discussion related to them carry over to the present situation. In fact, several of the publications referred to earlier have used equations of the form of Equations 11 and 12 and have displayed many of the transient features previously explored for the CSTR.

If one judges the adequacy of simplified versions of the problem, such as that described by Equations 11 and 12 or of other low-order lumping forms in terms of quantitative agreement with the original problem, he would conclude that they are inadequate in many situations (*see*, for example, such comparisons in the papers by Van den Bosch and Padmanabhan (*144*).) Remember, however, that the original equations are themselves highly idealized mathematical descriptions of a complicated underlying physical system. While they may be basically sound for reaction in a system of parallel pores of equal cross-section, length, and activity, there has been very little experimental testing of their adequacy in describing the quantitative behavior of a highly nonlinear reaction process in structures resembling typical commercial catalysts. Therefore, the original description itself should be viewed as useful mainly for qualitative purposes at present, and there is little justification for terming as inadequate the simplified lumped versions as long as they predict the same qualitative trends. From Equations 11 and 12, one would conclude, for example, by analogy to the CSTR problem that sustained oscillatory behavior is not possible as long as $Le > 1$ if $Sh = Nu$ or if $Nu, Sh \to \infty$; in these cases the slope condition is necessary and sufficient for stability determinations. They lead to the same conclusion for finite and unequal values of the Nusselt and Sherwood numbers unless $Nu > Sh$. These results are consistent with present knowledge of the behavior of solutions of Equations 3–6. Certain qualitative features of the solutions of the original distributed system, however, cannot be predicted by the simplest lumped form, and it is important that researchers or potential users of the simplified forms be aware of these. For example, such features discussed earlier as asymmetric steady profiles and a multiplicity greater than three for a first-order Arrhenius reaction require a higher-order approximation. It also appears that extremely steep concentration profiles are not easily described by these methods. For these, a two-zone model described by Paterson and Cresswell (*139*) appears to be a good alternative. In this model, a reaction zone through which the concentration of

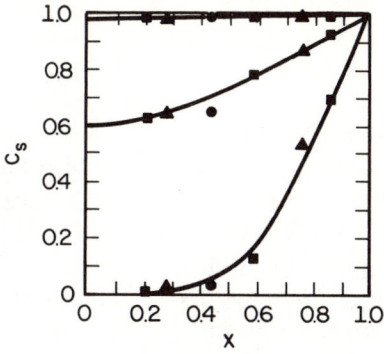

Chemical Engineering Science

Figure 8. Multiple steady-state concentration profiles in a catalyst slab obtained through numerical integration (———) and orthogonal collocation: ● *(N = 1),* ▲ *(N = 2),* ■ *(N = 3). Le = 1; $\gamma = 20$; $\beta' = 0.7$; $\phi = 0.16$; Nu = Sh = ∞ (140).*

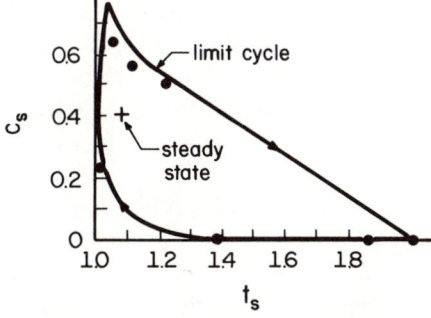

Figure 9. Sustained oscillations in a spherical particle obtained through orthogonal collocation (———) with N = 1 by Hellinckx et al. (140) and through numerical solution (●) by Lee and Luss (113). Le = 0.1; $\gamma = 30$; $\beta' = 0.15$; $\phi = 1.1$; Nu = Sh = ∞.

a limiting reagent decreases to zero is assumed to exist near the particle surface and is described mathematically by low-order collocation equations. Chemical reaction does not take place in the inner zone. Van den Bosch and Padmanabhan (144) found this technique to be useful in describing steep gradients and in predicting the results of Hatfield and Aris (mentioned previously) which show a multiplicity of five.

Evidence that qualitatively and quantitatively accurate information may be obtained in some cases with simple lumped models was shown well by Hellinckx *et al.* (*140*), from which Figures 8 and 9 were taken for illustration.

All in all, the results of studies of simplified models are encouraging. Little significant progress is imminent along the lines of extending our general knowledge to more complex chemical reactions or of putting our present knowledge to practical use in design work unless these or other simplifications can be proved useful.

Related Problems. Closely related to the problems of porous catalyst particles are those with reaction on external surfaces including catalytic wires and gauzes. The latter at least have long been of concern in industrial applications. In fact, the thrust of Liljenroth's paper in 1918 (*11*) was toward the ignition and extinction characteristics of ammonia oxidation on a catalytic gauze. In recent years, Luss and co-workers have studied various problems with catalytic wires including steady-state multiplicity and stability (*151*), temperature flickering (*152, 153, 154*), and standing wave solutions (*155*). Some of these include experimental data (*151, 153, 154*). Flickering, a cause of inefficient industrial operation, is apparently caused by concentration or mixing fluctuations in the impinging stream. Ray *et al.* (*156*) pointed out that the large thermal capacity of the wires and gauzes would most likely cause all oscillations arising from intrinsic instabilities to damp. Mathematical solutions which correspond to spatially standing waves along the wire length in cases of an infinite length or insulated ends are possible but have been shown to be unstable (*155, 157*).

The problem of steady-state multipilicity and stability with diffusion and conduction to a spherical catalytic surface has been studied in detail (*158, 159, 160*). A cognate subject is that of noncatalytic fluid-solid reactions. It is not possible here to probe the distinguishing characteristics of the mathematical models and the solutions for such systems. Discussions of them as well as literature references are in Refs. *161, 162, 163*.

Experimental Studies. There have been only a few published reports of experiments with single catalyst particles that have demonstrated steady-state multiplicity and/or instabilities. Those of which the author is aware are listed along with brief descriptions in Table II. Except for the study of carbon monoxide oxidation by Beusch *et al.* (first entry in the table), all of the studies involved large thermal effects, and the primary experimental measurement was of the temperature at the center of the catalyst particle. Commonly the experimental setup was one in which a gas stream containing the reactants flowed past a freely suspended catalyst particle by forced convection. In some of the experiments by Wicke's group, the particle was imbedded in a layer of inert particles.

Table II. Experimental Studies of Steady-State Multiplicity and Instabilities with Single Catalyst Particles[a]

Reference	Experimental System	o	m
1. Beusch, Fieguth, and Wicke, 1970, 1972 (*164,165*)	(a) Oxidation of hydrogen on suspended Pt/silica/alumina particle	x	x
	(b) Oxidation of hydrogen on single particle imbedded in a layer of inert particles	x	x
	(c) Oxidation of CO (nearly isothermal) on suspended Pt/Al$_2$O$_3$ pellet	x	x
2. Horak and Jiracek, 1970 (*166*)	Oxidation of hydrogen on a suspended Pt/Al$_2$O$_3$ catalyst particle		x
3. Furusawa and Kunii, 1971 (*167*)	Ethylene hydrogenation on a suspended pellet of Adkin's catalyst		x
4. Jiracek, Havlicek, and Horak, 1971 (*168*)	Oxidation of hydrogen on a Pt/alumina pellet		x
5. Horak and Jiracek, 1972 (*61*)	Oxidation of hydrogen on Pt/alumina particle in a backmix reactor and flow tubular reactor		x
6. Jiracek, Horak, and Hajkova, 1973 (*169*)	Oxidation of hydrogen on a Pt/alumina particle in a batch recycle-flow reactor		x

[a] Column headings are defined in Table I.

Although these experimental reports do not contain thorough quantitative comparisons with theoretical predictions, the qualitative features of the laboratory observations are readily appreciated in view of known theoretical behavior.

The careful experimental work of Wicke and his co-workers (*164, 165*) warrants special comment. Figures 10 and 11 demonstrate some of their results, including temperature oscillations with the hydrogen–oxygen reaction (Figure 10) and multiple steady states in experiments with carbon monoxide oxidation (Figure 11). A recent paper by Wicke (*170*) summarizes these and other contributions from his group at Münster. That oscillatory states exist is not surprising in view of aforementioned theoretical studies, but the oscillations observed in these experiments do not seem to be explainable in terms of interactions of transport processes —internal and external—and with a simple one-step kinetic mechanism of the type assumed in all theoretical studies of particle oscillations to date. Beusch *et al.* (*164, 165*) suggested that complicated surface kinetics, not reducible to a single rate expression, are involved. It is also possible that the adsorptive species capacitance of the catalytic surface, while not entering into the steady-state picture, affects the stability characteristics appreciably. This capacitance, which is unaccounted for in the preceding equations, is commonly overlooked.

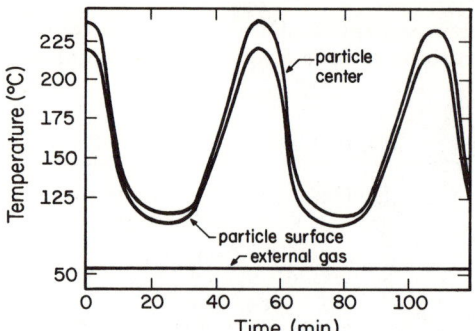

Figure 10. Temperature oscillations in a single 8-mm spherical catalyst particle (Pt/silica/alumina) during hydrogen oxidation in air (3.14 vol % H_2) (164, 165)

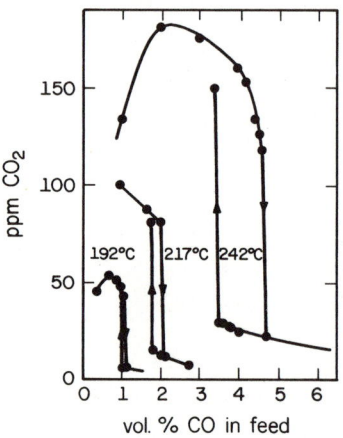

Figure 11. Multiple steady states (hysteresis) in the oxidation of CO in air in a single 3 × 3 mm cylindrical catalyst particle (Pt/Al_2O_3) (164, 165)

An experimental study by Benham and Denny (*171*) deserves comment because it includes quantitative comparisons of experimental measurements of unsteady temperature profiles with theoretical predictions. The experiments involved the oxidation of carbon monoxide on a cylindrical pellet of CuO/Al_2O_3 catalyst in forced convection flow. Seven thermocouples were imbedded in the pellet. The work was not aimed at studying steady-state multiplicity or instabilities; it is pertinent to this discussion because it demonstrates the difficulty and perhaps the accuracy which one might expect to find in attempting to predict catalyst particle behavior quantitatively. Benham and Denny point out that variations in transfer rates over the particle surface lead to markedly nonisothermal and asymmetric particle profiles. These measurements with a real catalyst particle therefore add emphasis to some of the comments above to the effect that the usual mathematical model given by Equations 3–6 should be considered useful primarily for qualitative purposes and that this should be remembered when simplified versions are evaluated.

In other experimental studies with a single catalyst particle, Kehoe and Butt (*172, 173*) demonstrated reasonable, though rough, agreement with results calculated from a model similar to that given by Equations 3 and 4. No steady-state multiplicity or instabilities were observed in those experiments. Certainly there is a need for further careful experimental work on these problems.

Tubular and Fixed-Bed Reactors

Mathematical Models and Theoretical Studies. Nearly all of the mathematical models which have been used in studies of multiple states, stability, and sensitivity of tubular or fixed-bed reactors use balance equations on the fluid phase of the following form or of forms readily deduced therefrom:

$$\frac{\partial c_f}{\partial \theta} = \frac{1}{Pe_c} \frac{\partial^2 c_f}{\partial \xi^2} - \frac{\partial c_f}{\partial \xi} - R_1 \tag{13}$$

$$L \frac{\partial t_f}{\partial \theta} = \frac{1}{Pe_t} \frac{\partial^2 t_f}{\partial \xi^2} - \frac{\partial t_f}{\partial \xi} + R_2 - \alpha(t_f - t_a) \tag{14}$$

The forms of the rate functions R_1 and R_2 depend on whether the model is "homogeneous" or "heterogeneous." With first-order Arrhenius-type kinetics, R_1 and R_2 for homogeneous models are:

$$R_1 = Da\, c_f \exp\left[\gamma \left(1 - \frac{1}{t_f}\right)\right] \tag{15}$$

$$R_2 = \beta R_1 \tag{16}$$

For the same kinetics, expressions for R_1 and R_2 with heterogeneous models are:

$$R_1 = \frac{\tau}{\tau'} \frac{1-\varepsilon}{\varepsilon} \varepsilon_s \phi^2 \eta\, c_f \exp\left[\gamma \left(1 - \frac{1}{t_f}\right)\right] \tag{17}$$

$$R_2 = \frac{\tau}{\tau'} (a+1) \frac{Nu}{Le} (L-1)(t_s - t_f) \tag{18}$$

The Danckwerts boundary conditions are commonly used, even in the unsteady state.

$$\xi = 0: \quad c_f = 1 + \frac{1}{Pe_c} \frac{\partial c_f}{\partial \xi}$$
$$t_f = 1 + \frac{1}{Pe_t} \frac{\partial t_f}{\partial \xi} \quad \quad (19)$$

$$\xi = 1: \quad \frac{\partial c_f}{\partial \xi} = \frac{\partial t_f}{\partial \xi} = 0 \quad \quad (20)$$

No lengthy discussion of the general mathematical model itself is intended here, but a few comments are warranted. (A recent paper by Froment (*174*) contains an extensive discussion of model representation of fixed-bed reactors. A recent paper by Lübeck (*175*) compares the dynamic behavior of the homogeneous model with that of several modifications of the two-phase model of a packed reactor within the region of steady-state multiplicity, all through numerical calculations. The influence of the boundary conditions was also investigated.) The homogeneous model is often adopted for studies of both packed (catalytic) reactors and empty tubular (truly homogeneous) reactors. For packed reactors the implicit assumption in the homogeneous model is that the solid catalytic material is always in thermal equilibrium with the fluid and that the chemical reaction, occurring either inside or on the external surface of catalyst particles, is unaffected by transport. An important quantity in this model of packed reactors is the capacity factor L. That factor may be so large that the concentration profile through the bed may be assumed to be in a pseudo-steady state relative to the instantaneous temperature profile. Studies in which L is taken to be unity properly apply only to an unpacked tubular reactor. Obviously L affects only the unsteady state.

As made evident by Equations 17 and 18, the heterogeneous model recognizes the presence of a catalyst phase and can account for transport effects both internal and external to the catalyst. For the heterogeneous model there must be included a set of equations, such as those in Equations 3–6, from which local and instantaneous values of η and t_s may be obtained.

A number of configurational variations, some of which are important in problems of interest here, such as partial recycle of the effluent (loop reactor) and co-current or countercurrent heat exchange with a coolant or with the reactor effluent, can easily be brought into the mathematical description by modifying the boundary conditions and/or attaching an additional energy balance to describe the variation of t_a.

Theoretical investigations into the multiplicity, stability, and sensitivity of states for many special cases deducible from the preceding equations have been reported. Most studies fall into one of three categories: (1) investigations of the effect of axial dispersion with a homogeneous model using Equations 13–16, 19, and 20 either with $\alpha = 0$ or t_a constant; (2) investigations of effects of the catalyst phase using a heterogeneous model with $Pe_c, Pe_t \to \infty$ and either $\alpha = 0$ or t_a constant in Equations 13, 14, and 17–20); and (3) investigations of the various special geometric configurations using a homogeneous model with $Pe_c, Pe_t \to \infty$. Consequently, most studies are simplified in the sense that they isolate one of the effects of dispersion, fluid-particle interactions, and reactor configuration from the others. The following discussion is organized somewhat along the lines of these three situations, but some others of interest are also included. Five of the six papers in Session VII of ADVANCES IN CHEMISTRY SERIES No. 133 deal with reactors of this type. An attempt is made here to put them in perspective.

Multiplicity of steady states in the homogeneous reactor model with axial mixing was first demonstrated by Van Heerden (8) by numerical solution of the governing steady-state equations. Raymond and Amundson (176) also presented a number of calculations and simulations, and a series of papers by Amundson and his co-workers (177, 178, 179, 180, 181, 182, 183, 184, 185) contains a complete and detailed mathematical study of this problem for a single reaction with $L = 1$. The most recent paper in that series (185) contains a number of interesting examples for which the stability of steady states was determined from computations of eigenvalues (by the Galerkin method) of the linearized equations, and results of computer simulations were shown for a nonadiabatic reactor. Various types of instabilities, resembling those of the CSTR, were demonstrated, including sustained oscillations.

The establishing of sufficient conditions for uniqueness (and stability in some cases) in terms of kinetic expressions and system parameters has been the continuing prey of theoreticians. These problems provide a good arena for applying such mathematical tools as fixed-point methods, comparison equation theorems, bifurcation theory, and Liapunov functionals. The review by Ray at the previous international symposium (2) describes some of these. More recently sufficient conditions for uniqueness have also been worked out by Matsuyama (186) for the adiabatic case with $Pe_c = Pe_t$ and by Han and Agrawal (187) and Varma and Amundson (183) for nonadiabatic cases with $Pe_c \neq Pe_t$. The common feature of all of these criteria, some of which are applicable only to a first-order Arrhenius reaction, is that they are satisfied if the Peclet numbers are large. Froment (188) pointed out that the values of the Peclet number (considering $Pe_c = Pe_t$) which would give rise to steady-state multiplicity and stability problems are at least an order of magnitude smaller than those applicable to industrial packed-bed reactors. He concluded that there is little likelihood of such problems being encountered except possibly in unusual cases of a very short reactor or short stages—as perhaps in the case described by Dassau and Wolfgang (73). Criteria for uniqueness, however, may not be so easily met in small experimental or pilot plant reactors. In this regard, recent papers by Hlavacek et al. (189) and Vortmeyer and Schaefer (190) are important. In the first, experimental evidence indicating that Pe_c is considerably greater than Pe_t was cited, and calculations showed that the region of steady-state multipilicity is extended to much higher values of Pe_c. The second paper showed that there is an equivalence of a sort between the homogeneous and heterogeneous models for cases of a packed reactor if the dispersion coefficients are suitably defined to reflect the transfer rates between the fluid and solid. The analysis and results of Vortmeyer and Schaefer do not apply in all cases—the major assumptions being that $\partial^2 t_f/\partial \xi^2 = \partial^2 t_s/\partial \xi^2$ and that the gas phase temperature is at a pseudo-steady state relative to the local instantaneous solid temperature; still they suggest that the behavior predicted by the homogeneous dispersion model may be more meaningful if Pe_c and Pe_t are not viewed in the strictest sense. The concluding statement by Varma and Amundson (185) to the effect that the interesting results which their analysis predicts beg for experimental confirmation is appropriate. Experimental confirmation is not yet available.

As one would expect intuitively and, in fact, can show formally, the homogeneous model reduces to the CSTR problem as the Peclet numbers become very small. Cohen (191) used singular perturbation methods to look at the small Peclet number case and described behavior analogous to that of the CSTR for $L = 1$. Hlavacek and Hofmann (192, 193) used the "linearization" method for first-order lumping (mentioned also in the previous section for the catalyst particle problem), also for $L = 1$, to study steady-state multiplicity, stability characteristics, and transient behavior, including sustained oscillations for small Peclet numbers. Their results compared favorably, in a qualitative sense at least, to numerical solutions. Low-order lumped models are probably less descriptive of real behavior here than in the catalyst particle problem, except for small Peclet numbers. Still one might hope that they would be useful in describing trends and parameter effects. According to the work of Hlavacek and Hofmann (193) with the first-order lumped model, sustained oscillatory states are not possible in an adiabatic reactor unless $Pe_t > Pe_c$. (This inequality can probably never be satisfied in a realistic situation where the thermal diffusivity would exceed molecular diffusivities because of radiation and solid conduction effects. Note that this inequality restriction does not apply to nonadiabatic reactors.)

Since the CSTR is a limiting case of the homogeneous tubular reactor, one expects the general behavioral features to be similar and, as in the catalyst particle problem, tends to describe any departure from CSTR behavior as a "surprise." One such feature, apparently first discovered in computations by McGowin and Perlmutter (*194*) and subsequently shown also by Hlavacek *et al.* (*197*) and Varma and Amundson (*184, 185*) is the existence of five steady states under some conditions (nonadiabatic in all cases reported) with a first-order Arrhenius reaction.

As mentioned earlier, virtually all studies of steady-state multiplicity in the homogeneous model with axial dispersion have been of a single exothermic reaction. A rare exception is the work of Chen (*198*) on a theoretical study of a simple first-order autocatalytic reaction describing isothermal microbial growth.

The important early contributions to the theory of the *heterogeneous* models were by Wicke and Vortmeyer (*199*), Wicke (*200*), Liu and Amundson (*201, 202*), and Liu, Aris, and Amundson (*203*). The analysis in these cases was based on a plug-flow model for the fluid phase with all diffusive transport effects confined to the fluid film surrounding the catalyst particles. Liu and Amundson (*202*) also considered the effect of axial dispersion in the fluid phase. None of these studies accounted for axial conduction from particle to particle so that each element of catalyst was effectively isolated from others. Investigations into the multiplicity and stability of steady states, therefore, were reduced to an examination of the behavior of each particle singly in surrounding conditions given by the local instantaneous state of the fluid. In some examples, these studies showed that over a portion of the bed, the state of the fluid would be such that the particles could have multiple states. In such situations, as long as there is no direct axial communication between the adjacent catalyst particles, the model allows for an unlimited number of steady-state profiles—if one constructs profiles composed of some particles in a high-temperature state and others in a low state through that portion of the reactor where multiplicity is possible. Except for a few of these, all obviously have discontinuous solid phase temperature profiles. More recently Aris and Schruben (*204*) studied this situation in detail using the Heaviside step function to represent the dependence of the heat generation rate on temperature. A further study was reported by Farina and Aris (*205*), who took into account fluid-phase dispersion and solid conduction (axial). The principal effect of the solid conduction is to smooth the discontinuous profiles and possibly reduce the total number of steady profiles to a small finite value.

The effects of axial conduction in the solid phase for a first-order Arrhenius reaction were demonstrated through numerical solutions of the unsteady equations by Eigenberger (*206*). He used fluid equations similar to Equations 13 and 14 with Pe_c, $Pe_t \to \infty$ and with R_1 given by Equation 18c; his equations for the solid phase were of the form of Equations 11 and 12 except that a term proportional to $\partial^2 t_s / \partial \xi^2$ was added to account for solid conduction. Concerning multiplicity of steady profiles, the essential result was that with adiabatic conditions imposed on the solid at the inlet and outlet boundaries, transient profiles would arrive at one of three different stable steady profiles; the one actually reached depended on initial conditions. Presumably two others lie between the stable ones for a multiplicity of five. Furthermore, the multiplicity was sensitive to inlet and outlet conditions on the solid energy equation. With radiation losses at the inlet, three stable steady states were still possible, but much less likely to occur, and if equations for packed sections of inert material at the inlet and outlet were appended, only two stable profiles were found. Eigenberger pointed out that the results of his simulations resemble the experimental findings of Padberg and Wicke (*207*). Since all of his results were obtained from numerical solutions for an example case, no definite conclusions regarding the possible number of steady states in such cases can be given, and the matter is still not completely resolved. Also unsettled are questions concerning the possibility of sustained oscillatory states and the stability of various steady profiles when axial conduction effects in the solid phase are taken into account.

In the work described above, Eigenberger also showed that the heterogeneous model accounting for axial conduction in the solid phase could describe "wandering" or "creeping" profiles. Such profiles (experimental data were apparently first given by Wicke and Vortmeyer (*199*)) are characterized by the occurrence of a steep temperature profile (ignition zone) over a small section of the reactor which alternately can be made to wander upstream or downstream slowly with slight parameter changes. According to Eigenberger's results, this phenomenon can be explained in terms of parametrically sensitive steady-state profiles; for these profiles the steady location of the ignition zone is changed drastically from one position to another in the reactor in response to a small change in a parameter. Large wall or solid heat capacities then give rise to the wandering nature of the profiles which are the slow transients accompanying the transition from one steady state to another. Other mathematical descriptions which have been used to explain these wandering profiles include those based on cell models by Vanderveen *et al.* (*208*) and Rhee *et al.* (*209*) and on a homogeneous model of reactors of infinite length by Vortmeyer and Jahnel (*210, 211*). So far, at least, experimental information on these curious transients (contained in the reference given above and in others listed later) has not discriminated between the various mathematical models which can describe them qualitatively.

With Pe_c, $Pe_t \to \infty$, Equations 13–16, 19, and 20 have a unique solution for given conditions if t_a is constant or with cocurrent flow of a coolant. Other reactor configurations, however, such as those with countercurrent cooling, heat exchange between effluent and feed, and with recycle streams, which provide a means of thermal feedback, can give ris to steady-state multiplicity and stability problems. Van Heerden (7, 8) and Bilous and Amundson (212) considered some of these cases. There have been a number of other theoretical studies of these problems (146, 191, 196, 213, 214, 215, 216, 217, 218, 219, 220, 221, 222, 223, 224, 225 226, 227, 228, 229, 230, 231, 232). In the recent literature these models have received relatively little attention, but multiplicity and stability problems caused by the effects contained in them may be much more frequently encountered in present industrial processes than by the effects of axial dispersion or reaction in porous catalyst particles. An example commonly referred to is the synthesis of ammonia in packed reactors with heat exchange between the reacting fluid and the feed (7, 214, 215, 229).

In the simple homogeneous plug flow reactor model without any mechanism of thermal feedback, severe design problems with reactor sensitivity may be encountered even though steady states are stable and unique. Sensitivity to small changes in a parameter or operating condition is very likely the explanation for most incidents of "runaway" conditions in industrial reactors. Barkelew (233) first presented some guidelines for designing reactors to avoid runaways caused by sensitivity. Other criteria for avoiding extremely sensitive conditions have been given by Dente and Collina (234), Hlavacek et al. (235), and Van Welsenaere and Froment (236). The agreement between these various

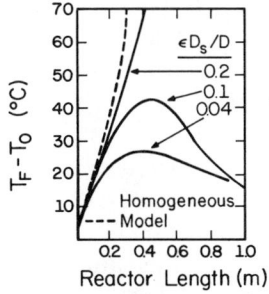

Chemical Engineering Science

Figure 12. Effect of intraparticle diffusion on the temperature profile in a one-dimensional packed-bed reactor with $Pe = \infty$ (120)

criteria for first-order Arrhenius kinetics is very good, as shown by Froment (*174*). Froment also showed the effect of various reactor models, including two-dimensional models, on the predicted reactor profiles and their sensitivity. There is no precise definition of sensitivity. That of Van Welsenaere and Froment (*236*), for example, was given in terms of the temperature at the hot spot and the existence of an inflection point in the temperature profile. The criteria established are intended to be used as guides in the first stages of reactor design. Reactors for which conditions lie close to the runaway region require individual study and simulations.

All of the criteria cited above for determining sensitivity or runaway conditions were based on homogeneous models. That these may be more conservative than necessary if diffusional limitations within (or external to) the catalyst particles are taken into account was illustrated by Cresswell and Paterson (*120*). They used the oxidation of o-xylene as an example. Figure 12 presents a comparison of temperature profiles for different values of the pore diffusivity for a case in which a homogeneous model predicts a runaway profile. Related studies by McGreavy and Cresswell (*237*) and McGreavy and Adderley (*238*) further emphasize this point and suggest that working with a more detailed model in designing reactors near the runaway limits or within the runaway region of homogeneous models may be well justified. McGreavy and Adderley (*127*) proposed a runaway limit which they obtained by examining the sensitivity of the particle temperature to the local external fluid temperature.

Experimental Studies. Table III lists experimental studies of steady-state multiplicity and stability in packed-bed reactors. As the number of entries indicates, there has been a paucity of experimental information relative to the amount of reported theoretical work on this type of reactor. [Some of the entries in Table I might have been placed in Table III because they used packed-bed reactors. However, those reactors were the recirculating type which behaved more nearly like CSTR's than tubular ractors.] The only report of oscillatory states is that of Volter (*239*) for the gaseous polymerization of ethylene. These oscillations were not linked quantitatively to a mathematical model. Entries 5, 7, 8 and 10 in the table represent studies of liquid phase reactions in empty tubes. Except for the experiments of Butakov and Maksimov (*246*), these have all been shown to follow the predictions of a plug flow model. Butakov and Maksimov used a nonadiabatic coiled tubular reactor. The behavior reported, shown in Figure 13, must be the result of an axial dispersion or wall conduction mechanism. No theoretical model was included in the paper.

Table III. Experimental Studies of Steady-State Multiplicity and Instabilities in Tubular and Fixed-Bed Reactors[a]

Reference	Experimetal System	o	m
1. Volter, 1963 (239)	Polymerization of ethylene in a nonadiabatic tubular reactor	x	
2. Padberg and Wicke, 1967 (207)	Oxidation of CO in fixed bed of Pt/Al_2O_3 catalyst		x
3. Wicke, Padberg, and Arens, 1968 (240)	Oxidation of ethane on Pd/alumina catalyst and oxidation of CO on Pt/alumina catalyst in a fixed bed		x
4. Kilger, 1969 (241)	Catalytic oxidation of carbon monoxide flowing longitudinally past a platinum tube		x
5. Root and Schmitz, 1969, 1970 (242, 243)	Liquid phase reaction between $Na_2S_2O_3$ and H_2O_2 in adiabatic tubular reactor with recycle		x
6. Fieguth and Wicke, 1971 (244)	Oxidation of CO in adiabatic fixed bed of Pt/Al_2O_3 catalyst		x
7. Luss and Medellin, (245)	Liquid phase reaction between $Na_2S_2O_3$ and H_2O_2 in nonadiabatic tubular reactor with countercurrent flow of coolant in an annulus		x
8. Butakov and Maksimov, 1973 (246)	Liquid phase polymerization of styrene in laminar flow nonadiabatic tubular reactor		x
9. Votruba and Hlavacek, 1973 (247)	Catalytic oxidation of carbon monoxide in adiabatic packed-bed reactors of Pd/Al_2O_3 and CuO/Al_2O_3 catalyst (three stable states observed)		x
10. Ausikaitis and Engel, 1974 (248)	Liquid phase reaction between $Na_2S_2O_3$ and H_2O_2 in a cycled batch reactor simulating a plug-flow tubular reactor with recycle		x

[a] Column headings are as defined in Table I.

Experiments reported with fixed-bed or catalytic reactors (entries 2–4, 6, and 9 in Table III) nicely demonstrate steady-state multipilicity and associated hysteresis phenomena. Carried out at low Peclet and Reynolds numbers, they apparently show effects of both axial dispersion and interphase transport. To some extent, they seem to be describable by more than one mathematical model, and close quantitative comparisons would be very difficult. Most of these experimental studies include some discussion of the associated theory. It is particularly noteworthy that Votruba and Hlavacek (entry 9 in Table III) observed three stable steady states under certain conditions. Prior theoretical studies, cited above, have shown this to be possible, but apparently it had not previously been observed experimentally.

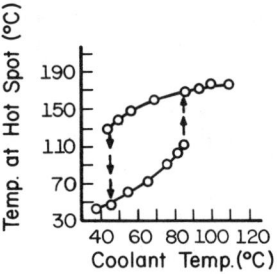

Doklady Akademii Nauk SSSR

Figure 13. Multiple steady states (hysteresis) in the liquid-phase polymerization of styrene in a tubular reactor (4 mm id coiled tube) in laminar flow (246)

In addition to the experiments on multiplicity and stability listed in Table III, there have been several reported experimental investigations of "wandering" profiles (*199, 207, 240, 249, 250, 251*)—a transient phenomenon described in the preceding section. As pointed out earlier, theoretical explanations have been offered in terms of different models, and experimental studies reported to date do not seem to distinguish among them.

Papers in Session VII. As mentioned earlier, five of the six papers in Session VII of the symposium volume apply directly to tubular and fixed-bed reactor dynamics. A brief critique is presented here as an appropriate follow-up of the preceding review of the existing literature on this subject. (The remaining paper in the session concerns more generally the subject of mixing. Though applicable in a general sense to reactors of interest here, it is more appropriately brought into the discussion in a later section.)

Two of the papers, one by Van Doesburg and de Jong and the other by Hansen and Jorgensen, study the unsteady behavior of fixed-bed reactors and compare theoretical predictions with experimentally measured profiles. Van Doesburg and de Jong studied the hydrogenation of CO and CO_2 to methane—parallel exothermic reactions—in an adiabatic reactor. They systematically consider each physical rate process and conclude that a simple plug-flow homogeneous model applies (with heat capacity of the solid material taken into account). Their model then consists of Equations 13 and 14, with $L > 1$ and $Pe_c, Pe_t \to \infty$, suitably modified to include more than one reaction. The results predicted by their model show very good agreement with their experimental data. In their experiments Pe_c varied from 300 to 1200, considerably beyond the value for which one would expect to encounter any appreciable effect of axial dispersion and, therefore, steady-state multiplicity.

In the experiments of Hansen and Jorgenson, the Peclet number was somewhat lower ($Pe_c \simeq 270$, $Pe_t \simeq 115$), and these workers used a homogeneous axial dispersion model to describe transients in the hydrogen–oxygen reaction in an adiabatic reactor packed with Pt/alumina catalyst. According to the calculations of Hlavacek et al. (189) for $Pe_c \neq Pe_t$, the value of Pe_c in these experiments is about twice the value at the upper limit of multiplicity. Hansen and Jorgenson do not report any observations of steady-state multiplicity. The authors point out that the experimentally observed transients in the last part of the reactor are not well described by their model. This observation suggests questions as to the adequacy of the dispersion model, particularly for describing the unsteady state. As in most real chemical systems, such discrepancies can be the result of an inaccurate representation of chemical rate processes, as the authors suggest.

Eigenberger describes an interesting theoretical study of the effects of the thermal capacitance and conductance of the reactor wall on the dynamic behavior in homogeneous plug-flow tubular reactors. Both a liquid reaction and a gas-phase reaction are considered, and the wall is not assumed to be in thermal equilibrium with the fluid. Two results are particularly interesting: (1) for a liquid reaction, longitudinal conduction in the wall is not significant; yet, owing to the heat capacity of the wall, some features of the dynamic response are similar to the wandering profiles described earlier for fixed beds; (2) for a gas phase reaction, wall conductivity effects are important and lead to multiple steady states under some conditions. A recent publication by Eigenberger (252) gives additional attention to wall effects on tubular reactor dynamics.

Hlavacek and Votruba report some interesting results on the oxidation of CO. Their experiments involved various catalysts in fixed beds and in monolithic honeycomb structures. Such reactors have been discussed extensively earlier in this symposium in the review paper on catalytic mufflers by Wei and are also the principal topic of a paper in Session IX by Young and Finlayson. They differ from the usual commercial catalytic reactor in that the Peclet and Reynolds numbers are small (laminar flow). Hlavacek and Votruba report experimental observations of multiple steady states in both types of reactors. The observation of multiple states in the flow of reactants through channels with catalytic walls (as in the honeycomb structure) does not seem to have been reported previously, though the experiments by Kilger cited in Table III were similar. This particular observation by Hlavacek and Votruba is interesting in view of the theoretical results of Young and Finlayson; the latter authors show essentially that while a simple one-dimensional heterogeneous model incorporating constant heat and mass transfer coefficients can have multiple steady solutions, a more exact two-dimensional solution using a fully developed velocity distribution is always unique. Young and Finlayson did not account for longitudinal heat conduction in the reactor wall, but they point out that this may be important in some cases. The theoretical results of Young and Finlayson as well as those of Eigenberger (cited above, regarding wall effects), suggest that wall conduction (which in effect is an axial dispersion mechanism) is the physical process responsible for multiplicity in the experiments with the honeycomb structure. Hlavacek and Votruba do not discuss two-dimensional models but consider several variations of the general one-dimensional model given above.

The paper in this session by McGreavy and Adderley focuses on the problem of designing fixed-bed reactors to avoid parametric sensitivity and multiple steady states. Early work on the sensitivity problem dealt entirely with homogeneous models and more recent work by McGreavy and co-workers emphasized possible importance of transport limitations in this regard. In their paper at this symposium, McGreavy and Adderley extend this notion to obtain criteria in graphical form in terms of operating parameters. The results permit a designer to determine whether operating parameters are such that the reactor will be relatively insensitive to small parameter changes and will have unique states. It is further suggested that these results might be readily implemented in the computer control or optimization of a given reactor.

Mixing and Modeling—Effects on Multiplicity and Stability

By heuristic arguments, Van Heerden (8) advanced the notion that a necessary condition for multiplicity of steady states in an exothermic reacting system was that some physical mechanism for heat feedback to exist. He demonstrated this point by working out examples of a CSTR, a plug-flow tubular reactor with feed-effluent heat exchange, and a tubular reactor with axial dispersion. Obviously the general requirement of feedback should not be restricted to exothermic systems but must apply to accelerating reactions in general. The notion has served a useful purpose although no precise definition of feedback has been offered. The notion is interwoven in much of the following discussion of mixing effects and modeling considerations—a discussion appropriately brief and limited to intrinsic hydrodynamic effects.

Clearly in order for intrinsic feedback in a given reacting flow system to be possible, the residence times must be distributed so that fluid elements of different ages can intermix. Residence-time distribution information, however, describes only macroscopic mixing features and by itself tells nothing about mixing at the microscopic level—*i.e.*, about the state of segregation as described by Danckwerts (*253*). If fluid elements in any reacting flow situation are completely segregated with regard to both heat and mass exchange, there can be no feedback regardless of the macroscopic picture. (Mathematically, completely segregated systems give rise to problems of the initial-value type.) Thus both macroscopic and microscopic effects warrant consideration. Nevertheless, almost all mathematical models which have been used to study multiplicity, stability, and sensitivity of states are based on macroscopic considerations. In the axial dispersion model of a tubular or fixed-bed reactor, for example, the effective axial diffusivity can be chosen to describe macroscopic effects, and once it is chosen, the level of microscopic mixing is also fixed.

In Session VII Yang and Weinstein describe a way to study theoretically both macroscopic and microscopic effects. They maintain that two adjustable parameters, n and R, in a hypothetical reactor model consisting of n CSTR's in series and a recycle stream (with recycle ratio R) can be separately and simultaneously manipulated so that the model can describe given residence-time distribution data and still provide flexibility in covering a range of micromixing conditions between complete segregaton and maximum mixedness—the extreme mixing situations described by Zwietering (*254*). Furthermore, the model is quite amenable to mathematical analysis and seems well suited for empirical reactor modeling in many practical situations. Yang and Weinstein present results of calculations showing the effect of micromixing on steady-state multiplicity in an exothermic reaction.

Apparently the only prior study of the effects of micromixing on multiplicity and stability was that by Yamazaki and Ichikawa (*255*); they examined the extreme cases of complete segregation of both heat and mass and of maximum mixedness for an arbitrary residence-time distribution. Their conclusion was that steady states in the first case are always unique and stable and that those in the second are stable if a perfectly mixed reactor having the same mean residence time as the given reactor is stable.

The discussion of modeling and the notion of feedback in connection with steady-state multiplicity and instabilities can be pursued further along somewhat different lines. Admittedly any real chemically reacting system has an underlying mathematical description which in its exact form —if an exact form can even be constructed—is far too complex to analyze. Simplifications and approximations must be made. Thought-provoking questions then arise as to how one can be reasonably certain that the simplifications and approximations he introduces do not also introduce spurious qualitative features into the solutions of the mathematical model —features that have no counterpart either in the real physical system or in the solutions of the more exact mathematical problem. (Simplifications, of course, may also eliminate important qualitative features.) There probably is no general answer to such questions, but as a reasonable guideline, it appears that one should be wary whenever an approximation changes the basic mathematical form. Some examples are readily available. One is in the aforementioned paper by Young and Finlayson in Session IX. Laminar flow through a catalytic duct with negligible axial diffusion or conduction gives rise to a nonlinear parabolic system of equations when formulated properly in terms of radial and axial position variables. For this problem, as in most parabolic problems, the solutions are unique and stable—no feedback mechanism exists. If, however, as Young and Finlayson show, the problem is simplified to a one-dimensional form in which radial variations are accounted for only in terms of constant heat and mass transfer coefficients to the catalytic duct wall, uniqueness can no longer be guaranteed. In fact, this simplified version may have an infinite number of solutions as discussed in some publications cited earlier. The real physical situation may indeed have non-unique solutions (the experiments described in the paper by Hlavacek and Votruba in Session VII show this to be true), but these must be the result of axial wall conduction (or of molecular conduction and diffusion axially in the tube) and not solely of transport between the catalytic wall and the fluid in laminar flow. The use of constant transfer coefficients is valid only for uniformly accessible surfaces. Catalyst particles in a packed bed may provide areas that are nearly uniformly accessible because of the nature of the flow.

In a paper addressed to these same matters, Lindberg and Schmitz (256) considered a theoretical problem of flow past a nonuniformly accessible catalytic surface—namely boundary layer flow past wedge-shaped solids. The conclusion was similar to that of Young and Finlayson; the solution of the boundary layer problem was shown to be unique, but the solution of a simplified version using constant heat and mass transfer coefficients was multivalued. Lindberg and Schmitz included a discussion of the general modeling problem and of the possible pitfalls that one would hope to avoid.

Consider, as another example, fully developed constant-property laminar flow in a tubular reactor with homogeneous exothermic reaction and negligible axial conduction and diffusion of heat and mass both in the fluid and in the wall. The two-dimensional steady-state material and energy balances are of the parabolic form (an initial-boundary value problem), and solutions are unique for realistic reaction kinetic models. No mechanism of thermal feedback exists. If the notion of Taylor diffusion is introduced (to account for dispersion effects arising from radial velocity gradients), then the model is converted to the one-dimensional axial dispersion form, a nonlinear boundary value system. A feedback mechanism has been introduced through the simplification procedure, and the uniqueness and stability of steady-state solutions can no longer be guaranteed. In light of these comments, the experimental observations of Butakov and Maksimov (*see* Table III), which involved the laminar flow of liquid reactants with no obvious intrinsic mechanism of feedback other than those of molecular diffusion and conduction, are surprising.

All such considerations should remind researchers not to lose sight of the underlying physical problem when attacking the mathematical one. Clearly they also must be kept in mind if experimental observations are to be carefully and correctly interpreted.

Other Areas of Application

Steady-state multiplicity, instabilities, and oscillatory behavior in reacting systems are also of interest in biology, combustion, and electrochemistry. In each of these areas a large literature has developed—all in parallel for the most part with few points of contact. From a mathematical viewpoint, at least, problems in these areas closely resemble those connected with chemical reactor design. It seems appropriate, therefore, to describe some of the research and the interesting problems encompassed in these areas, even though the coverage must be brief and inadequate.

In biological applications the major interest has been in oscillating systems. The pioneering work in this regard was by Lotka (*257, 258*) in the early 1900's on the oscillations in predator–prey interactions. Since then, oscillatory behavior at all levels of biological activity has been observed and studied. It is frequently suggested in these studies that intrinsic instabilities in biochemical reactions are responsible for circadian (daily) and other rhythms so prevalent in living systems. A recent review by Nicolis and Portnow (*259*) covers chemical oscillators with particular regard to biological applications. A number of books have been devoted to biological rhythms; a recent one by Pavlidis (*260*) emphasizes the mathematical analysis of them.

Steady-state multipilicity has been invoked in biological problems to explain switching and threshold phenomena in biochemical pathways and the developmental processes whereby a developed organism composed of many different types of cells (different "final" states) emanates from a single, nearly uniform fertilized egg cell (*see,* for example, the book by Rosen (*261*) and papers by Edelstein (*262*) and Lavenda (*263*).)

Explanations for the development of form and structure in living systems has also been offered in terms of spatial or "symmetry breaking" instabilities in systems involving diffusion and reaction. This phenomena was studied first by Turing (*264*) and later by Prigogine and co-workers

(*265, 266, 267, 268*) and Scriven and co-workers (*269, 270*) (*see* also the books by Glansdorff and Prigogine (*271*) and Aris (*5*)). In such theoretical studies, questions are posed as to whether spatially dependent perturbations, imposed on an initially uniform steady state, grow (or decay) uniformly or instead grow in time with a certain preferred spatial structure. Under certain conditions a spatially periodic growth is predicted. Evidence that such instabilities actually occur in biological systems may be found in various descriptions of pattern and aggregative movement of cells in culture such as those described by Elsdale (*272*) and Bonner (*273*) (*see* also the aforementioned book by Pavlidis).

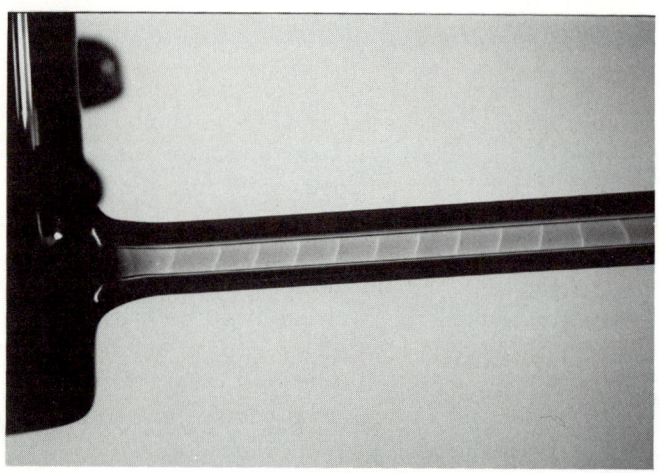

Chemical Engineering Communications

Figure 14. Traveling chemical waves in a diffusion tube during the Belousov-Zhabotinskii reaction (with Fe^{3+}). The light bands (waves) are blue; the dark regions are red. Waves, initiated by oscillations in the stirred beaker (at left of photo), are about 4 mm apart and are traveling at a speed of about 0.5 cm/min (280).

Most mathematical models used in studies related to biological applications are for isothermal kinetics, and they allow for the exchange of material with the surroundings, or "openness," by invoking the assumption that some species concentrations have fixed constant values. The resulting mathematical descriptions are similar to those of the usual continuous-flow chemical reactor models, but the underlying physical picture differs. The resulting steady states are actually pseudo-steady states if the system is truly closed, or they represent steady states in extreme cases of open systems for which some components can be exchanged with the surroundings without resistance.

Though not a biological reaction, the isothermal liquid-phase Belousov-Zhabotinskii reaction has attracted great interest among biologists and biophysicists as well as chemists mainly because the intriguing behavior which it displays is reminiscent of biological behavior. (Refs. *19* and *23*, which contain studies of this reaction in an open, well-mixed system, have been cited in an earlier section.) Oscillations, indicated by sharp color changes from red to blue (when Fe^{3+} is used as the metal ion) with the reaction proceeding in a closed stirred beaker, were described by Zhabotinskii (*274*). Spatially periodic behavior, visible as an assembly of travelling waves, in the absence of stirring, were later reported by Busse (*275*) and Zaiken and Zhabotinskii (*276*). A detailed physical description of the form and interactions of these travelling waves has been given by Winfree (*277, 278, 279*), the latter reference containing some beautiful color photographs of spiralling waves. The propagation of these waves through a small diffusion tube was studied experimentally by Tatterson and Hudson (*280*) (*see* Figure 14). A thorough study of this system by Field *et al.* (*33*) revealed that it consists of 10 coupled reactions including an autocatalytic sequence. At present there is no proof that the observed spatially periodic phenomena are the result of spatial or "symmetry breaking" instabilities of the type studied theoretically in references cited above. Rather the waves which propagate through stagnant mixtures seem to emanate from boundary perturbations, such as dust or minute particles. A general theoretical analysis of the effects of perturbations of this type has been given by Ortoleva and Ross (*281*).

In other studies motivated by biological applications, Aris and Keller (*282*) and Bailey and Luss (*283*) have suggested that asymmetric concentration profiles through membranes, similar to those described earlier for catalyst particles, resulting from the possibility of multiple states in enzymatic reactions may have a bearing on active transport.

A very interesting paper in the combustion literature by Gray *et al.* (*284*) crosses the boundaries of combustion, chemical reactor, and biological applications. Among the various topics included and analyzed are hyperthemia in warm-blooded animals—described as a thermal runaway when the body temperature exceeds a critical value—and hibernation which is explained in terms of the form of the heat generation curves for hibernating animals which permit multiple states, the two stable ones being the hibernating state and the active state.

Most of the theory of multiple states, instabilities, and sensitivity for chemical reactors can be applied to combustion problems. In combustible systems, multiple steady states in open processes are the rule rather than the exception; the region of multiplicity corersponds to conditions over which the mixture can be brought either to a steady ignited state or to an extinguished one by appropriate perturbations or start-up conditions. Standard references for the treatment and discussion of these and related topics are the books by Frank-Kamenetski (285) and Vulis (286). Much of the theory and experimental information pertain to closed systems—i.e., to situations in which reactants are charged to a vessel, subjected to a predetermined pressure and ambient temperature, and observed through an ensuing transient. Even for such cases, however, the theory usually is applied to a "pseudo" steady-state version of the problem, and when gradients are neglected, the results resemble those described earlier in this paper for the CSTR (see, for example, a recent review by Berlad (287). Longwell and Weiss (288) introduced the continuous-flow, well-stirred combustor, similar to a CSTR, as a convenient experimental tool and demonstrated its utility by using extinction data—i.e., the limit of steady-state multiplicity corresponding to a transition from a high temperature state to a low one—to deduce kinetic parameters. Similar theoretical and experimental studies in distributed-parameter flow systems have also been reported (289, 290, 291).

Instabilities and multiple steady states occur in electrochemical systems because of highly nonlinear current-potential relationships. A description by Wojtowicz (292) is a good reference source. In it methods of Liapunov and Poincaré are used to analyze dynamics, and behavioral features are described which have analogs in chemical reactor behavior. According to this book the first cases of periodic behavior in electrochemical systems were reported as early as 1828. Alkire and Nicolaides (293, 294) recently discovered that the equations governing a certain distributed model of the aeration corrosion of a metal under a moist film possess at least 13 steady-state profiles for the current distribution. Furthermore, some of these lead to very highly localized reaction rates and suggest a possible explanation or mechanism for the localized corrosive attack of metals.

Problems of intrinsic instabilities arise in many other situations which do not involve chemical rate processes. It is not possible to describe these in detail here, but they are worthy of note because the mathematical methods used to study them and the phenomena of interest are often very similar to those described here for chemical systems. Among these are problems in hydrodynamics and crystallization. Denn (6) treats the former and provides some unification of these problems with those of chemical reactors. The occurrence of sustained oscillations in perfectly mixed crystallizers was studied from a theoretical viewpoint by Sherwin et al. in 1967 (295) and more recently by Randolph et al. (296). Experimental studies of these oscillations have also been reported (297, 298).

Concluding Remarks

One noticeable recent trend has been the increase in experimental information. Until very recently there was a dearth of experimental data or reports of actual chemical reactor behavior to support the large number of theoretical studies. Many theoretical results have now been borne out by experiments, and laboratory data definitely add a tone of realism to problems in this area. Still, theory leads experimental fact by a considerable measure although some experimental observations have suggested, and probably will continue to suggest, questions for further theoretical investigations. The need for additional laboratory studies is most evident in distributed problems where some viewpoints need clarification and where predictions of the usual mathematical models could be qualitatively incorrect.

From a theoretical viewpoint, at least, the problem of elucidating the possible behavioral features in systems of reactions (without having to handle each case individually) is a challenging one—even for the lumped CSTR case. Until now, the vast majority of work has been with a single exothermic reaction. Emphasis has been on the effects of interacting physical rate processes. Some of the experimental observations of oscillations in single catalyst particles and of the fascinating conduct of the Belousov-Zhabotinskii system, described in studies cited above, as well as applications in some key areas of biology, may stimulate more effort in the direction of complex reacting systems. Only a very small percentage of the papers being published on the subjects of this review are from industry. As a result, the present practical applicability of academic research to industrial reactor design and its impact thereon are uncertain.

Acknowledgments

Helpful comments were obtained from Dan Luss, James Douglas, Gerhart Eigenberger, and John Villadsen, who kindly reviewed a preliminary version of this paper.

Nomenclature

a	geometric constant; 0 for slab, 1 for cylinder, 2 for sphere
a_v	area for heat losses per unit reactor volume
b	characteristic length; half-width for slab, radius of cylinder or sphere
c_f	dimensionless concentration of reactant in the bulk fluid phase, C_{Af}/C_{AO}
c_s	dimensionless concentration of reactant in particle pores, C_{Af}/C_{AO}
C_A	concentration of reactant
C_p	heat capacity

D	binary diffusion coefficient
D_f	effective axial diffusivity of reactant
D_s	pore diffusivity of reactant
Da	Damkohler number given by $K\tau \exp(-\gamma)$ or $K_o\tau$
E	activation energy
F	volumetric flow rate
h	fluid-particle heat transfer coefficient based on superficial external particle surface area
$(-\Delta H)$	heat of reaction (positive for exothermic reaction)
k_f	effective axial fluid thermal conductivity
k_m	fluid-particle mass transfer coefficient based on superficial external particle surface area
k_s	effective thermal conductivity of catalyst particle
K	pre-exponential factor for reaction rates per unit of reactor volume V
K'	pre-exponential factor for reaction rates per unit of pore volume
K_o, K_o'	reaction velocity constants at temperature T_o, given by $K \exp(-\gamma)$ and $K' \exp(-\gamma)$, respectively
l	tubular reactor length
L	dimensionless capacity factor, $1 + \dfrac{1-\epsilon}{\epsilon} \dfrac{\rho_s C_{ps}}{\rho_f C_{pf}}$
Le	Lewis number, $\rho_s C_{ps} D_s / k_s$
m	mass
n_1, n_2	lumping constants
N	number of spatial positions (e.g., collocation points) in lumping procedure
Nu	Nusselt number (Biot number), hb/k_s
Pe_c	Peclet number for mass dispersion, vl/D_f
Pe_t	Peclet number for heat dispersion, $vlc_{pf}\rho_f/k_f$
R_1, R_2	rate functions
R_g	universal gas constant
Sh	Sherwood number, $k_m b / \epsilon_s D_s$
t_f	dimensionless temperature of bulk fluid phase, T_f/T_o
t_s	dimensionless temperature within catalyst particle, T_s/T_o
T	temperature
U	overall coefficient for heat losses from the reactor
v	average interstitial fluid velocity
V	void reactor volume
x	dimensionless distance, y'/b
y	distance variable measured from reactor inlet
y'	distance variable measured from center of catalyst particle
z	time
α	dimensionless coefficient for reactor heat losses $Ua_v\tau/\rho_f C_{pf}$
β	dimensionless adiabatic temperature rise in fluid phase, $(-\Delta H)C_{AO}/\rho_f C_{pf} T_o$
β'	dimensionless adiabatic temperature rise in catalyst particle, $(-\Delta H)\epsilon_s D_s C_o / T_o k_s$
γ	dimensionless activation energy, $E/R_g T_o$
ϵ	reactor void fraction, ratio of bulk fluid volume to total volume
ϵ_s	porosity of catalyst particle

η	effectiveness factor for catalyst particle relative to reaction rate at C_{Af}, T_f
θ	dimensionless time for CSTR and tubular reactors, z/τ
θ'	dimensionless time for catalyst particle, z/τ'
λ_c, λ_t	lumping constants defined in Equation 10
ξ	dimensionless distance, y/l
ρ	density
τ	characteristic time for CSTR and tubular reactors, V/F
τ'	characteristic time for catalyst particle, b^2/D_s
ϕ	Thiele modulus, $b\sqrt{K_o'/D_s}$

Subscripts

a	ambient or coolant conditions
f	bulk fluid phase
o	feed conditions
s	state within catalyst particle or submerged solid material

Literature Cited

1. Gavalas, G. R., "Nonlinear Differential Equations of Chemically Reacting Systems," Springer-Verlag, New York, 1968.
2. Ray, W. H., *5th European/2nd Int. Symp. Chem. Reaction Eng., Amsterdam, 1972*, p. A8-1, (1972).
3. Perlmutter, D. D., "Stability of Chemical Reactors," Prentice-Hall, Englewood Cliffs, 1972.
4. Oppelt, W., Wicke, E., Eds., "Grundlagen der Chemischen Prozessregelung," Oldenbourg Verlag, Munich, 1964.
5. Aris, R., "The Mathematical Theory of Diffusion and Reaction in Permeable Catalysts," Vols. I and II, Clarendon Press, Oxford, 1975.
6. Denn, M. M., "Stability of Reaction and Transport Processes," Prentice-Hall, Englewood Cliffs, 1975.
7. Van Heerden, C., *Ind. Eng. Chem.* (1953) **45**, 1242.
8. Van Heerden, C., *Chem. Eng. Sci.* (1958) **8**, 133.
9. Bilous, O., Amundson, N. R., *AIChE J.* (1955) **1**, 513.
10. Aris, R., Amundson, N. R., *Chem. Eng. Sci.* (1958) **7**, 121, 132, 148.
11. Liljenroth, F. G., *Chem. Met. Eng.* (1918) **19**, 287.
12. Wagner, C., *Chem. Techn.* (1945) **18**, 28.
13. Poore, A. B., *Arch. Rational Mech. Anal.* (1973) **52**, 358.
14. Uppal, A., Ray, W. H., Poore, A. B., *Chem. Eng. Sci.* (1974) **29**, 967.
15. Gilles, E. D., Hofmann, H., *Chem. Eng. Sci.* (1961) **15**, 328.
16. Schmitz, R. A., Amundson, N. R., *Chem. Eng. Sci.* (1963) **18**, 265, 391, 415.
17. Skryabin, B. N., *Dokl. Akad. Nauk SSSR* (1968) **179**, 400.
18. Hlavacek, V., Kubicek, M., Jelinek, J., *Chem. Eng. Sci.* (1970) **25**, 1441.
19. Othmer, H., *Math. Biosciences*, in press.
20. Shastry, J. S., Fan, L. T., *Chem. Eng. J.* (1973) **6**, 129.
21. Heberling, P. V., Gaitonde, N. Y., Douglas, J. M., *AIChE J.* (1971) **17**, 1506.
22. Rayzak, R. J., Luus, R., *AIChE J.* (1971) **17**, 435.
23. Luus, R., Lapidus, L., *AIChE J.* (1972) **18**, 1060.
24. Douglas, J. M., Rippin, D. W. T., *Chem. Eng. Sci.* (1966) **21**, 305.
25. Bailey, J. E., *Chem. Eng. Commun.* (1973) **1**, 111.
26. Goldstein, R. P., Amundson, N. R., *Chem. Eng. Sci.* (1965) **20**, 195, 449, 477, 501.
27. Elnashaie, S., Yates, J. G., *Chem. Eng. Sci.* (1973) **28**, 515.

28. Bukur, D. B., Wittmann, C. V., Amundson, N. R., *Chem. Eng. Sci.* (1974) **29**, 1173.
29. Newbold, F. R., Amundson, N. R., *Chem. Eng. Sci.* (1973) **28**, 1731.
30. Hlavacek, V., Kubicek, M., Visnak, K., *Chem. Eng. Sci.* (1972) **27**, 719.
31. Sabo, D. S., Dranoff, J. S., *AIChE J.* **16**, 211.
32. Graziani, K. R., Hudson, J. L., Schmitz, R. A., *66th Ann. Meetg. AIChE, Philadelphia, 1973; Joint GVC-AIChE Meetg., Munich, 1974.*
33. Field, R. J., Körös, E., Noyes, R. M., *J. Amer. Chem. Soc.* (1972) **94**, 8649.
34. Aris, R., Amundson, N. R., *Chem. Eng. Progr.* (1957) **53** (5), 227.
35. Nemanic, D. J., Tierney, J. W., Aris, R., Amundson, N. R., *Chem. Eng. Sci.* (1959) **11**, 199.
36. Hoftyzer, P. J., Zwietering, Th. N., *Chem. Eng. Sci.* (1961) **14**, 251.
37. Demo, H. R., Iglesias, O. A., Farina, I. H., Willis, E., DeSantiago, M., *Chem. Eng. Sci.* (1972) **27**, 2151.
38. Hyun, J. C., Aris, R., *Chem. Eng. Sci.* (1972) **27**, 1341, 1361.
39. Luyben, W. L., *AIChE J.* (1974) **20**, 175.
40. Higgins, J., *Ind. Eng. Chem.* (1967) **59** (5), 19.
41. Gross, B., Kim, Y. G., *Bull. Math. Biophys.* (1969) **31**, 441.
42. O'Neill, S. P., Lilly, M. D., Rowe, P. N., *Chem. Eng. Sci.* (1971) **26**, 173.
43. Chi, C. T., Howell, J. A., Pawlowsky, U., *Chem. Eng. Sci.* (1974) **29**, 207.
44. Matsuura, T., Kato, M., *Chem. Eng. Sci.* (1967) **22**, 171.
45. Knorr, R. S., O'Driscoll, K. F., *J. Appl. Polymer Sci.* (1970) **14**, 2683.
46. Horn, F., *Arch. Rational Mech. Anal.* (1972) **49**, 172.
47. Horn, F., Jackson, R., *Arch. Rational. Mech. Anal.* (1972) **47**, 81.
48. Feinberg, M., Horn, F. J. M., *Chem. Eng. Sci.* (1974) **29**, 775.
49. Horn, F., *Proc. Roy. Soc. Lond.* (1973) **A334**, 299, 313, 331.
50. Hofmann, H., *3rd European Symp. Chem. Reaction Eng., Amsterdam, 1964,* p. 235 (1965).
51. Hafke, C., Gilles, E. D., *Mess. Steuern Reg.* (1968) **11**, 204.
52. Hugo, P., *4th European Symp. Chem. Reaction Eng., Brussels, 1968,* p. 459 (1971).
53. Bush, S. F., *Proc. Roy. Soc.* (1969) **A309**, 1.
54. Furusawa, T., Nishimura, H., Miyauchi, T., *J. Chem. Eng. Japan* (1969) **2**, 95.
55. Baccaro, G. P., Gaitonde, N. Y., Douglas, J. M., *AIChE J.* (1970) **16**, 249.
56. Hugo, P., *Ber. Bunsen-Gesellschaft* (1970) **74**, 121.
57. Vejtasa, S. A., Schmitz, R. A., *AIChE J.* (1970) **16**, 410.
58. Gerrens, H., Kuchner, K., Ley, G., *Chem. Ing. Tech.* (1971) **43**, 693.
59. Horak, J., Jiracek, F., Krausova, L., *Chem. Eng. Sci.* (1971) **26**, 1.
60. Hancock, M. D., Kenney, C. N., *5th European/2nd Int. Symp. Chem. Reaction Eng., Amsterdam, 1972,* p. B3-47 (1972).
61. Horak, J., Jiracek, F., *5th European/2nd Int. Symp. Chem. Reaction Eng., Amsterdam, 1972,* p. B8-1 (1972).
62. Hugo, P., Jakubith, M., *Chem. Ing. Tech.* (1972) **44**, 383.
63. Lo, S. N., Cholette, A., *Can. J. Chem. Eng.* (1972) **50**, 71.
64. Dubil, H., Gaube, J., *Chem. Ing. Tech.* (1973) **45**, 529.
65. Eckert, E., Hlavacek, V., Marek, M., *Chem. Eng. Commun.* (1973) **1**, 89.
66. Eckert, E., Hlavacek, V., Kubicek, M., Sinkule, J., *Chem. Ing. Tech.* (1973) **45**, 83.
67. Marek, M., "Scientific Papers of Institute of Chemical Technology, Prague, in press.
68. Chang, M., Schmitz, R. A., *Chem. Eng. Sci.* (1975) **30**, 21 (second paper will appear in same journal).
69. Ding, J. S. Y., Sharma, S., Luss, D., *Ind. Eng. Chem., Fundamentals* (1974) **13**, 76.
70. Guha, B. K., Agnew, J. B., *Ind. Eng. Chem. Proc. Des. Dev.,* in press.

71. Weiss, A. H., John, T., *J. Catal.* (1974) **32**, 216.
72. Westerterp, K. R., *Chem. Eng. Sci.* (1962) **17**, 423.
73. Dassau, W. J., Wolfang, G. H., *Chem. Eng. Progr.* (1963) **59** (4), 43.
74. Cowley, P. E. A., Johnson, D. E., *ISA J.* (Feb. 1965) 81.
75. Herbrich, D., Thiele, R., Lucas, K., *Chem. Tech.* (1968) **20**, 75.
76. Iscol, L., *Proc. JACC* (1970), paper 23-13.
77. Schmeal, W. R., Barkelew, C. H., *63rd Ann. Meetg. AIChE, Chicago, 1970*, paper 23d.
78. Lee, W., Kugelman, A. M., *Ind. Eng. Chem., Product Res. Dev.* (1973) **12**, 197.
79. Hlavacek, V., Kubicek, M., Marek, M., *J. Catal.* (1969) **15**, 17.
80. *Ibid.*, p. 31.
81. Aris, R., *Chem. Eng. Sci.* (1969) **24**, 149.
82. Spalding, D. B., *Chem. Eng. Sci.* (1959) **11**, 53.
83. Kuo, J. C. W., Amundson, N. R., *Chem. Eng. Sci.* (1967) **22**, 49.
84. *Ibid.*, p. 443.
85. *Ibid.*, p. 1185.
86. Elnashaie, S. S. E. H., Cresswell, D. L., *Can. J. Chem. Eng.* (1973) **51**, 201.
87. Weisz, P. B., Hicks, J. S., *Chem. Eng. Sci.* (1962) **17**, 265.
88. Ostergaard, K., *Chem. Eng. Sci.* (1963) **18**, 259.
89. Amundson, N. R., Raymond, L. R., *AIChE J.* (1965) **11**, 339.
90. Luss, D., *Chem. Eng. Sci.* (1971) **26**, 1713.
91. Jackson, R., *Chem. Eng. Sci.* (1972) **27**, 2205.
92. Kastenberg, W. E., *Chem. Eng. Sci.* (1973) **28**, 1691.
93. Liou, C. T., Lim, H. C., Weigand, W. A., *AIChE J.* (1973) **19**, 321.
94. Copelowitz, I., Aris, R., *Chem. Eng. Sci.* (1970) **25**, 885.
95. *Ibid.*, p. 906.
96. Michelsen, M. L., Villadsen, J., *Chem. Eng. Sci.* (1972) **27**, 751.
97. Hlavacek, V., Marek, M., *Chem. Eng. Sci.* (1968) **23**, 865.
98. Gel'fand, I. M., *Amer. Math. Soc. Transl.* (1963) **29**, 295.
99. Fujita, H., *Bull. Amer. Math. Soc.* (1969) **75**, 132.
100. Jackson, R., *Chem. Eng. Sci.* (1973) **28**, 1355.
101. Hatfield, B., Aris, R., *Chem. Eng. Sci.* (1969) **24**, 1213.
102. Pis'men, L. M., Kharats, Y. I., *Dokl. Akad. Nauk SSSR* (1968) **178**, 901.
103. Horn, F. J. M., Jackson, R., Martel, E., Patel, C., *Chem. Eng. J.* (1970) **1**, 79.
104. Bailey, J. E., *Chem. Eng. Sci.* (1972) **27**, 1055.
105. Jackson, R., Horn, F. J. M., *Chem. Eng. J.* (1972) **3**, 83.
106. Luss, D., Bailey, J. E., Sharma, S., *Chem. Eng. Sci.* (1972) **27**, 1555.
107. Patel, C., Jackson, R., *5th European/2nd Int. Symp. Chem. Reaction Eng., Amsterdam, 1972*, p. 135-49 (1972).
108. Yang, R. Y. K., Padmanabhan, L., Lapidus, L., *Chem. Eng. J.* (1972) **2**, 218.
109. Yang, R. Y. K., Lapidus, L., *Chem. Eng. Sci.* (1973) **28**, 875.
110. Wei, J., *Chem. Eng. Sci.* (1965) **20**, 729.
111. Gavalas, G. R., *Chem. Eng. Sci.* (1966) **21**, 477.
112. Hlavacek, V., Marek, M., *4th European Symp. Chem. Reaction Eng., 1968*, p. 107 (1971).
113. Lee, J. C. M., Luss, D., *AIChE J.* (1970) **16**, 621.
114. Yang, R. Y. K., Lapidus, L., *Chem. Eng. Sci.* (1974) **29**, 1567.
115. Hlavacek, V., Kubicek, M., *Chem. Eng. Sci.* (1970) **25**, 1527.
116. Lih, M. M., Lin, K., *AIChE J.* (1973) **19**, 832.
117. Jackson, R., *Chem. Eng. Sci.* (1974) **29**, 1413.
118. Hansen, K. W., *Chem. Eng. Sci.* (1971) **26**, 1555.
119. Cresswell, D. L., *Chem. Eng. Sci.* (1970) **25**, 267.
120. Cresswell, D. L., Paterson, W. R., *Chem. Eng. Sci.* (1970) **25**, 1405.

121. McGreavy, C., Thornton, J. M., ADVAN. CHEM. SER. (1972) **109**, 607.
122. McGreavy, C., Thornton, J. M., *Chem. Eng. Sci.* (1970) **25**, 303.
123. McGreavy, C., Thornton, J. M., *Can. J. Chem. Eng.* (1970) **48**, 187.
124. McGreavy, C., Thornton, J. M., *Chem. Eng. J.* (1970) **1**, 296.
125. McGreavy, C., Soliman, M. A., *Chem. Eng. Sci.* (1973) **28**, 1401.
126. McGreavy, C., Thornton, J. M., *Chem. Eng. J.* (1973) **6**, 91.
127. McGreavy, C., Adderley, C. I., *Chem. Eng. Sci.* (1973) **28**, 577.
128. Hartman, J. S., Roberts, G. W., Satterfield, C. N., *Ind. Eng. Chem., Fundamentals* (1967) **6**, 80.
129. Mitschka, P., Schneider, P., *Coll. Czech. Chem. Comm.* (1970) **35**, 1617.
130. *Ibid.* (1971) **36**, 54.
131. Mehta, B. N., Aris, R., *J. Math. Anal. Appl.* (1971) **36**, 611.
132. Mehta, B. N., Aris, R., *Chem. Eng. Sci.* (1971) **26**, 1699.
133. Villadsen, J. V., Stewart, W. E., *Chem. Eng. Sci.* (1967) **22**, 1483.
134. Finlayson, B. A., "The Method of Weighted Residuals and Variational Principles," Academic Press, New York, 1972.
135. Stewart, W. E., Villadsen, J. V., *AIChE J.* (1969) **15**, 28.
136. Ferguson, N. B., Finlayson, B. A., *Chem. Eng. J.* (1970) **1**, 327.
137. Hlavacek, V., Kubicek, M., *J. Catal.* (1971) **22**, 364.
138. McGowin, C. R., Perlmutter, D. D., *Chem. Eng. Sci.* (1971) **26**, 275.
139. Paterson, W. R., Cresswell, D. L., *Chem. Eng. Sci.* (1971) **26**, 605.
140. Hellinckx, L., Grootjans, J., Van Den Bosch, B., *Chem. Eng. Sci.* (1972) **27**, 644.
141. Michelsen, M. L., Villadsen, J., *Chem. Eng. Sci.* (1972) **27**, 751.
142. Sorensen, J. P., Guertin, E. W., Stewart, W. E., *AIChE J.* (1973) **19**, 969.
143. Karanth, N. G., Koh, H. P., Hughes, R., *Chem. Eng. Sci.* (1974) **29**, 451.
144. Van Den Bosch, B., Padmanabhan, L., *Chem. Eng. Sci.* (1974) **29**, 805, 1217.
145. Hlavacek, V., Marek, M., Kubicek, M., *Coll. Czech. Chem. Comm.* (1970) **35**, 2124.
146. Hlavacek, V., Kubicek, M., Caha, J., *Chem. Eng. Sci.* (1971) **26**, 1743.
147. Hlavacek, V., Kubicek, M., *Chem. Eng. Sci.* (1971) **26**, 1737.
148. Mitschke, P., Schneider, P., *Coll. Czech. Chem. Comm.* (1972) **37**, 196, 2506.
149. Luss, D., Lee, J. C. M., *Chem. Eng. Sci.* (1971) **26**, 1433.
150. Lee, J. C. M., Padmanabhan, L., Lapidus, L., *Ind. Eng. Chem., Fundamentals* (1972) **11**, 117.
151. Cardoso, M. A. A., Luss, D., *Chem. Eng. Sci.* (1969) **24**, 1699.
152. Ervin, M. A., Luss, D., *Chem. Eng. Sci.* (1972) **27**, 339.
153. Edwards, W. M., Worley, Jr., F. L., Luss, D., *Chem. Eng. Sci.* (1973) **28**, 1479.
154. Edwards, W. M., Zuniga-Chaves, J. E., Worley, Jr., F. L., Luss, D., *AIChE J.* (1974) **20**, 571.
155. Luss, D., Ervin, M. A., *Chem. Eng. Sci.* (1972) **27**, 315.
156. Ray, W. H., Uppal, A., Poore, A. B., *Chem. Eng. Sci.* (1974) **29**, 1330.
157. Jackson, R., *Chem. Eng. Sci.* (1972) **27**, 2304.
158. Petersen, E. E., Friedly, J. C., De Vogelaere, R. J., *Chem. Eng. Sci.* (1964) **19**, 683.
159. Friedly, J. C., Petersen, E. E., *Chem. Eng. Sci.* (1964) **19**, 783.
160. Winegardner, D. K., Schmitz, R. A., *AIChE J.* (1968) **14**, 301.
161. Ishida, M., Wen, C. Y., *Chem. Eng. Sci.* (1971) **26**, 1043.
162. Wang, S. C., Wen, C. Y., *AIChE J.* (1972) **18**, 1231.
163. Knapp, R. H., Aris, R., *Arch. Rational Mech. Anal.* (1972) **44**, 165.
164. Beusch, H., Fieguth, P., Wicke, E., ADVAN. CHEM. SER. (1972) **109**, 615.
165. Beusch, H., Fieguth, P., Wicke, E., *Chem. Ing. Tech.* (1972) **44**, 445.
166. Horak, J., Jiracek, F., *Chem. Tech.* (1970) **22**, 393.
167. Furusawa, T., Kunii, D., *J. Chem. Eng. Japan* (1971) **4**, 274.

168. Jiracek, F., Havlicek, M., Horak, J., *Coll. Czech. Chem. Comm.* (1971) **36**, 64.
169. Jiracek, F., Horak, J., Hajkova, M., *Coll. Czech. Chem. Comm.* (1973) **38**, 185.
170. Wicke, E., *Chem. Ing. Tech.* (1974) **46**, 365.
171. Benham, C. B., Denny, V. E., *Chem. Eng. Sci.* (1972) **27**, 2163.
172. Kehoe, J. P. G., Butt, J. B., *AIChE J.* (1972) **18**, 347.
173. Kehoe, J. P. G., Butt, J. B., *5th European/2nd Int. Symp. Chem. Reaction Eng., Amsterdam, 1972*, p. B8-13 (1972).
174. Froment, G. F., *Chem. Ing. Tech.* (1974) **46**, 374.
175. Lubeck, B., *Chem. Eng. J.* (1974) **7**, 29.
176. Raymond, L. R., Amundson, N. R., *Can. J. Chem. Eng.* (1964) **48**, 173.
177. Amundson, N. R., *Can. J. Chem. Eng.* (1965) **43**, 49.
178. Luss, D., Amundson, N. R., *Chem. Eng. Sci.* (1967) **22**, 253.
179. Luss, D., Amundson, N. R., *Can. J. Chem. Eng.* (1967) **45**, 341.
180. *Ibid.* (1968) **46**, 424.
181. Markus, L., Amundson, N. R., *J. Diff. Eqns.* (1968) **4**, 102.
182. Varma, A., Amundson, N. R., *Chem. Eng. Sci.* (1972) **27**, 907.
183. Varma, A., Amundson, N. R., *Can. J. Chem. Eng.* (1972) **50**, 470.
184. *Ibid.* (1973) **51**, 206.
185. *Ibid.*, p. 459.
186. Matsuyama, H., *J. Chem. Eng. Japan* (1973) **6**, 276.
187. Han, C. D., Agrawal, S., *Chem. Eng. Sci.* (1973) **28**, 1617.
188. Froment, G. F., ADVAN. CHEM. SER. (1972) **109**, 1.
189. Hlavacek, V., Hofmann, H., Votruba, J., Kubicek, M., *Chem. Eng. Sci.* (1973) **28**, 1897.
190. Vortmeyer, D., Schaefer, R. J., *Chem. Eng. Sci.* (1974) **29**, 485.
191. Cohen, D. A., *Summer Inst. Nonlinear Math. Battelle, Seattle, 1972*.
192. Hlavacek, V., Hofmann, H., *Chem. Eng. Sci.* (1970) **25**, 173.
193. *Ibid.*, p. 1517.
194. McGowin, C. R., Perlmutter, D. D., *AIChE J.* (1971) **17**, 831.
195. *Ibid.*, p. 837.
196. *Ibid.*, p. 842.
197. Hlavacek, V., Hofmann, H., Kubicek, M., *Chem. Eng. Sci.* (1971) **26**, 1629.
198. Chen, M. S. K., *AIChE J.* (1972) **18**, 849.
199. Wicke, E., Vortmeyer, D., *Z. Elektrochem.* (1959) **63**, 145.
200. Wicke, E., *Z. Elektrochem.* (1961) **65**, 267.
201. Liu, S. L., Amundson, N. R., *Ind. Eng. Chem., Fundamentals* (1962) **1**, 200.
202. *Ibid.* (1963) **2**, 183.
203. Liu, S. L., Aris, R., Amundson, N. R., *Ind. Eng. Chem., Fundamentals* (1963) **2**, 12.
204. Aris, R., Schruben, D. L., *Chem. Eng. J.* (1971) **2**, 179.
205. Farina, I. H., Aris, R., *Chem. Eng. J.* (1972) **4**, 149.
206. Eigenberger, G., *Chem. Eng. Sci.* (1972) **27**, 1909, 1917.
207. Padberg, G., Wicke, E., *Chem. Eng. Sci.* (1967) **22**, 1035.
208. Vanderveen, J. W., Luss, D., Amundson, N. R., *AIChE J.* (1968) **14**, 636.
209. Rhee, H. K., Foley, D., Amundson, N. R., *Chem. Eng. Sci.* (1973) **28**, 607.
210. Vortmeyer, D., Jahnel, W., *Chem. Ing. Tech.* (1971) **43**, 461.
211. Vortmeyer, D., Jahnel, W., *Chem. Eng. Sci.* (1972) **27**, 1485.
212. Bilous, O., Amundson, N. R., *AIChE J.* (1956) **2**, 117.
213. Douglas, J. M., Orcutt, J. C., Berthiaume, P. W., *Ind. Eng. Chem., Fundamentals* (1962) **1**, 253.
214. Baddour, R. F., Brian, P. L. T., Loglais, B. A., Eymery, J. P., *Chem. Eng. Sci.* (1965) **20**, 281.

215. Brian, P. L. T., Baddour, R. F., Eymery, J. P., *Chem. Eng. Sci.* (1965) **20**, 297.
216. Reilly, M. J., Schmitz, R. A., *AIChE J.* (1966) **12**, 153.
217. *Ibid.* (1967) **13**, 519.
218. Wang, F. S., Perlmutter, D. D., *Proc. JACC, Philadelphia, 1967*, p. 107.
219. Wang, F. S., Perlmutter, D. D., *AIChE J.* (1968) **14**, 976.
220. Inoue, H., Komiya, T., *Int. Chem. Eng.* (1968) **8**, 749.
221. Mukosei, V. I., Pis'men, L. M., Kharkats, Y. I., *Int. Chem. Eng.* (1968) **8**, 17.
222. Pareja, G., Reilly, M. J., *Ind. Eng. Chem., Fundamentals* (1969) **8**, 442.
223. Pis'men, L. M., *Fiz. Goreniya i Vzryva* (1969) **5**, 239.
224. Matsuyama, H., *J. Chem. Eng. Japan* (1970) **3**, 248.
225. McGowin, C. R., Perlmutter, D. D., *Chem. Eng. J.* (1971) **2**, 125.
226. Matsuyama, H., *J. Chem. Eng. Japan* (1972) **5**, 197.
227. Volin, Y-M., Ostrovskii, G. M., Slin'ko, M. G., *Dokl. Akad. Nauk SSSR* (1972) **204**, 648.
228. Caha, J., Hlavacek, V., Kubicek, M., *Chem. Ing. Tech.* (1973) **45**, 1308.
229. Stephens, A. D., Richards, R. J., *Automatica* (1973) **9**, 65.
230. Liou, C. T., Weigand, W. A., Lim, H. C., *Int. J. Control* (1974) **19**, 561.
231. Liou, C. T., Lim, H. C., Weigand, W. A., *Chem. Eng. Sci.* (1974) **29**, 705.
232. Oh, S. H., Schmitz, R. A., *Chem. Eng. Comm.* (1974) **1**, 199.
233. Barkelew, C. H., *Chem. Eng. Progr. Symp. Ser.* (1959) **55** (25), 38.
234. Dente, M., Collina, A., *Chim. Ind.* (1964) **46**, 1445.
235. Hlavacek, V., Marek, M., John, T. M., *Coll. Czech. Chem. Comm.* (1969) **34**, 3868.
236. Van Welsenaere, R. J., Froment, G. F., *Chem. Eng. Sci.* (1970) **25**, 1503.
237. McGreavy, C., Cresswell, D. L., *4th European Symp. Chem. Reaction Eng., Brussels, 1968*, p. 59 (1971).
238. McGreavy, C., Adderley, C. I., *Chem. Eng. Sci.* (1974) **29**, 1054.
239. Volter, B. V., *Proc. Int. Fed. Automatic Control Congr., 1963*, p. 507/1.
240. Wicke, E., Padberg, G., Arens, H., *4th European Symp. Chem. Reaction Eng., Brussels, 1968*, p. 425 (1971).
241. Kilger, H., *Chem. Ing. Tech.* (1969) **41**, 862.
242. Root, R. B., Schmitz, R. A., *AIChE J.* (1969) **15**, 670.
243. *Ibid.* (1970) **16**, 356.
244. Fieguth, P., Wicke, E., *Chem. Ing. Tech.* (1971) **43**, 604.
245. Luss, D., Medellin, P., *5th European/2nd Int. Symp. Chem. Reaction Eng., Amsterdam, 1972*, p. 134 (1972).
246. Butakov, A. A., Maksimov, E. I., *Dokl. Akad. Nauk SSSR* (1973) **209**, 643.
247. Votruba, J., Hlavacek, V., *Chem. Prunysl* (1973) **11**, 541 [English translation in *Int. Chem. Eng.* (1974) **14**, 461].
248. Ausikaitis, J., Engel, A. J., *AIChE J.* (1974) **20**, 256.
249. Vortmeyer, V. D., *Z. Elektrochem.* (1961) **65**, 282.
250. Berty, J. M., Bricker, J. H., Clark, S. W., *5th European/2nd Int. Symp. Chem. Reaction Eng., Amsterdam, 1972*, p. B8-27 (1972).
251. Venkatachalam, P., Kershenbaum, L., Grossmann, E., Earp, R., *5th European/2nd Int. Symp. Chem. Reaction Eng., Amsterdam, 1972*, p. B8-39 (1972).
252. Eigenberger, G., *Chem. Ing. Tech.* (1974) **46**, 11.
253. Danckwerts, P. V., *Chem. Eng. Sci.* (1958) **8**, 93.
254. Zwietering, Th. N., *Chem. Eng. Sci.* (1959) **11**, 1.
255. Yamazaki, H., Ichikawa, A., *J. Chem. Eng. Japan* (1969) **2**, 100.
256. Lindberg, R. C., Schmitz, R. A., *Chem. Eng. Sci.* (1969) **24**, 1113.
257. Lotka, A., *J. Phys. Chem.* (1910) **14**, 271.
258. Lotka, A., *Proc. Nat. Acad. Sci. U.S.* (1920) **6**, 410.
259. Nicolis, G., Portnow, J., *Chem. Rev.* (1973) **73**, 365.

260. Pavlidis, T., "Biological Oscillators: Their Mathematical Analysis," Academic Press, New York, 1973.
261. Rosen, R., "Dynamical System Theory in Biology," Wiley-Interscience, New York, 1970.
262. Edelstein, B. B., *J. Theor. Biol.* (1972) **37**, 221.
263. Lavenda, B. H., *Quart. Rev. of Biophys.* (1972) **5** (4), 429.
264. Turing, A. M., *Phil. Trans. of the Roy. Soc.* (1952) **B237**, 37.
265. Prigogine, I., Nicolis, G., *J. Chem. Phys.* (1967) **46**, 3542.
266. Prigogine, I., Lefever, R., *J. Chem. Phys.* (1968) **48**, 1695.
267. Prigogine, I., Lefever, R., Goldbeter, A., Herschkowitz-Kaufman, M., *Nature* (1969) **233**, 913.
268. Prigogine, I., Babloyantz, A., "Analysis and Simulation of Biological Systems," H. C. Hemker and B. Hess, Eds., American Elsevier, New York, 1972.
269. Gmitro, J. I., Scriven, L. E., "Intracellular Transport," K. B. Warren, Ed., Academic Press, New York, 1966.
270. Othmer, H. C., Scriven, L. E., *J. Theor. Biol.* (1971) **32**, 507.
271. Glansdorff, P., Prigogine, I., "Thermodynamic Theory of Structure, Stability and Fluctuations," Wiley-Interscience, New York, 1971.
272. Elsdale, T., "Towards a Theoretical Biology," C. H. Waddington, Ed., Vol. 4, p. 95, 1972.
273. Bonner, J. T., *Ann. Rev. Microbiol.*, C. E. Clifton, Ed., (1971) **25**, 75.
274. Zhabotinskii, A. M., *Dokl. Akad. Nauk SSSR* (1964) **157**, 392.
275. Busse, H. G., *J. Phys. Chem.* (1969) **73**, 750.
276. Zaikin, A. N., Zhabotinskii, A. M., *Nature* (1970) **225**, 535.
277. Winfree, A. T., *Science* (1972) **175**, 634.
278. *Ibid.* (1972) **181**, 937.
279. Winfree, A. T., *Sci. Am.* (1974) **230** (6), 82.
280. Tatterson, D. F., Hudson, J. L., *Chem. Eng. Commun.* (1973) **1**, 3.
281. Ortoleva, P., Ross, J., *J. Chem. Phys.* (1973) **58**, 5673.
282. Aris, R., Keller, K. H., *Proc. Nat. Acad. Sci.* (1972) **69**, 777.
283. Bailey, J. E., Luss, D., *Proc. Nat. Acad. Sci.* (1972) **69**, 1460.
284. Gray, B. R., Gray, P., Kirwan, N. A., *Combust. Flame* (1972) **18**, 439.
285. Frank-Kamenetskii, D. A., "Diffusion and Heat Transfer in Chemical Kinetics," 2nd ed., Plenum Press, New York, 1961.
286. Vulis, L. A., "Thermal Regimes of Combustion," McGraw-Hill, New York, 1961.
287. Berlad, A. L., *Combust. Flame* (1973) **21**, 275.
288. Longwell, J. P., Weiss, M. A., *Ind. Eng. Chem.* (1955) **47**, 1634.
289. Smith, H. W., Schmitz, R. A., Ladd, R. G., *Comb. Sci. Tech.* (1971) **4**, 131.
290. Fang, M., Schmitz, R. A., Ladd, R. G., *Comb. Sci. Tech.* (1971) **4**, 143.
291. Williams, F. A., *Ann. Rev. Fluid Mech.* (1971) **3**, 171.
292. Wojtowicz, J., "Modern Aspects of Electrochemistry," J. O'M. Bockris and B. E. Conway, Eds., Vol. 8, p. 47, 1972.
293. Alkire, R., Nicolaides, G., *J. Electrochem. Soc.* (1974) **121**, 183.
294. Alkire, R., Nicolaides, G., *J. Electrochem. Soc.* (1975) **122**, 25.
295. Sherwin, M. B., Shinnar, R., Katz, S., *AIChE J.* (1967) **13**, 1141.
296. Randolph, A. D., Beer, G. L., Keener, J. P., *AIChE J.* (1973) **19**, 1140.
297. Randolph, A. D., Beckman, J., 67th Ann. Meetg. AIChE, Washington, 1974, paper 100b.
298. Douglas, J. M., Song, Y. H., 67th Ann. Meetg. AIChE, Washington, 1974, paper 100c.

RECEIVED December 4, 1974. Work supported by the National Science Foundation and the Army Research Office, Durham, N. C.

8

Oxidation Reaction Engineering

A. CAPPELLI

Montedison Fibre R&D, Milan, and Istituto di Chimica Industriale, Politecnico di Milano, Italy

> *This review presents a selection of work on various aspects of oxidation reactions. The papers reviewed are divided into four sections: (a) vapor phase oxidation with oxygen or air, (b) liquid phase oxidation with oxygen, (c) liquid phase oxidation with hydroperoxides, and (d) heterogeneization of homogeneous catalytic processes. The papers of the first (and most important) section are examined according to process type and certain basic aspects. Information is given about the state of the art of oxidation reactors. Special attention is paid to automotive exhaust treatment processes. When the data are evaluated from the point of view of chemical reaction engineering, it is concluded that communication between researchers and process engineers needs improvement.*

The following review is a representative selection of work on various aspects of oxidation reactions published in 1972, 1973, and 1974. Three general sections are discussed which correspond to a natural division of the subject matter. These are: (a) vapor phase oxidation with oxygen or air, (b) liquid phase oxidation with oxygen, and (c) liquid phase oxidation with hydroperoxides. A fourth section, on the heterogeneization of homogeneous processes, in which the preparation of supported homogeneous catalysts is treated, is discussed separately. Special emphasis is given to automotive exhaust.

Vapor Phase Oxidation with Oxygen or Air

The discussion focuses on heterogeneous catalytic processes, which are by far the most important ones. Papers are reviewed according to different aspects which are of particular interest in this area. First considered are the types of processes, which include the oxidation of organic and some inorganic molecules. Secondly, some of the basic chemical and/or physical aspects of oxidation, such as mechanisms, nature of active centers, reaction patterns, effects of operating conditions and catalyst modifications, and role of diffusion are considered. Finally, some information about the present state of the art in the field of oxidation reactors, both on a laboratory and industrial scale, is given.

Types of Processes. In the papers examined the oxidation of olefins, aromatics, NH_3, SO_2, and CO have been studied. With regard to the oxidation of olefins, the reactions studied are listed in Table I, approximately according to their degree of oxidation.

DEHYDRODIMERIZATION. Trimm *et al.* (*1*) attempt to identify a catalyst for the polymerization and subsequent cyclization of olefins under oxidative conditions, using the oxidation of propene to benzene as an illustration. A survey of possible compounds showed that indium oxide could be a suitable catalyst. The oxide was tested experimentally and found to be a selective catalyst for the oxidation of propene to benzene. 1,5-Hexadiene and acrolein were produced in the early stages of the reaction, and the diene oxidized further to produce benzene. The kinetics of the reaction were examined in some detail, and a tentative mechanism was advanced.

The note by Parera *et al.* (*2*) reports the analogous oxidation of isobutene to 2,5-dimethyl-1,5-hexadiene and *p*-xylene over indium oxide. The kinetics were examined, and a mechanism was proposed.

OXIDATIVE DEHYDROGENATION. Sterrett *et al.* (*3*) present a kinetic study of the oxidative dehydrogenation of butene to butadiene over a zinc chromium–iron catalyst. The data on which the kinetic model is based were obtained using a set of statistically designed experiments. Selectivity to butadiene remained high throughout the runs. The formation of butadiene was fit to a semiempirical rate expression by a non-linear, least-square, curve-fitting technique.

In the study reported by Pitzer (*48*) methods of activating a phosphorus–tin oxide catalyst, active and selective for the oxidative dehydrogenation of butenes to butadiene, were investigated. Objectives of activation included several properties of the catalyst, and the author attempted to accomplish these objectives by heating the finished catalyst in steam, air, and nitrogen. Catalytic activity was improved by steaming at elevated temperatures while heating in air and nitrogen gave no improvement. Only macroporosity seemed to account for the increase in catalytic activity, and steaming appeared to affect the bulk of the catalyst instead of altering only the surface of the catalyst particles.

ALLYLIC OXIDATION. Five papers describe experiments carried out over molybdate catalysts. In the study by Wragg *et al.* (*5*) the ammoxidation of propene and of acrolein was studied over two catalysts—one

Table I. Vapor Phase

Process	Catalyst
Dehydromerization	indium oxide
↓	
Oxidative dehydrogenation	zinc–chromium oxide
	phosphorus–tin oxide
↓	molybdate
	tungstate
Allylic oxidation	uranium–antimonium oxide
	mercuric chloride
↓	silver
	molybdate
Oxygen insertion	(Pd doped) vanadium pentoxide
	molybden–alumina
↓	
Oxidative C–C cleavage	supported iridium
↓	
Complete combustion	platinum–alumina spinels
	chromite

similar to commercial bismuth molybdate catalysts and the other consisting of the koechlinite phase Bi_2O_3–MoO_3. The rates, reaction orders, Arrhenius parameters, and selectivities of the ammoxidations were studied at a fixed temperature, and tentative mechanisms have been advanced. Mann and Ko (6) report the effect of several variables on conversion and yield of the oxidation of 2-methylpropene to methacrolein over a bismuth molybdate catalyst. A rate equation has been evaluated based on a mechanism. Pasquon et al. (7) investigated the catalytic behavior of some molybdate catalysts in the oxidation of 1-butene to butadiene or to maleic anhydride. The catalytic action of bismuth molybdates is discussed by Schuit (8) in its connection with the solid structure, method of preparation, kinetics of the reaction catalyzed, and adsorption of reactants and products. Mechanisms for the oxidation of olefins and for the ammoxidation of propene and ammonia are proposed. Daniel and Keulks (9) report preliminary results of an investigation on a catalyst containing Bi, Mo, and Fe. The authors have found that the activity and selectivity of this catalyst are comparable with bismuth molybdate for the oxidation of propene to acrolein. Villa et al. (10) report the results of an investigation on the catalytic behavior of Bi tungstates for the oxidation and ammoxidation of olefins (propene and 1-butene). According to this study Bi_2WO_6 is the only active and selective compound. In a laboratory study by Grasselli and Suresh (4) a uranium–antimony oxide catalyst, known to be particularly efficient for synthesizing acrylonitrile, was studied to develop an understanding of structural features related to

Oxidation of Olefins

Reference	Examples
1, 2	propene → benzene
3	
48	butene → butadiene
5, 6, 7, 8, 9	
10	propene → acrolein
4	butene → butadiene
16	propene → acrylonitrile, acrylic acid
11, 29, 30, 31	ethylene → ethylene oxide
7, 12, 13	butene → maleic anhydride
14	ethylene → acetaldehyde
15	propene → acetone
25	propene → acetaldehyde
17, 18	hydrocarbons and CO → CO_2
19	

catalytic activity. A mechanism for the oxidation and ammoxidation of propene is proposed involving allylic intermediates. Arai *et al.* (*16*) presents a kinetic study of the oxidation of isobutene to methacrolein over mercuric chloride supported on active charcoal. The kinetics of the oxidation were determined, and reaction rates of other olefins were also measured.

OXYGEN INSERTION. Metcalf *et al.* (*11*) studied the kinetics of silver-catalyzed ethylene oxidation. The behavior of various inhibitors was investigated, and the kinetic data were fitted to Langmuir-Hinselwood rate expressions, although some inconsistencies were noted. Marcinkowsky *et al.* (*29*), Carberry *et al.* (*30*) and Forzatti *et al.* (*31*) also report studies on ethylene oxidation over Ag supported catalysts.

Trifirò *et al.* (*12*) present a study of the oxidation of 1-butene to maleic anhydride over a Mn–MoO_6 based catalyst. A monocenter oxidation mechanism, accounting for the formation of CO, CO_2, and maleic anhydride, is proposed. In the above mentioned paper by Pasquon *et al.* (*7*) on the catalytic behavior of some molybdate catalysts a tentative mechanism of formation of maleic anhydride from 1-butene over Fe–MoO_4 is advanced. Akimoto *et al.* (*13*) report the results of an investigation on supported molybdena catalysts for the oxidation of butadiene. Evnin *et al.* (*14*) describe work on the development of a heterogeneous catalytic system, consisting of palladium-doped vanadium pentoxide and a third component, which is capable of oxidizing ethylene directly to acetaldehyde with high specificity, activity, and stability. The results of an investigation on the catalytic activity of a $MoO_3 \cdot Al_2O_3$ system for the oxidation of propene are reported by Giordano et al. (15), and a tentative mechanism is advanced.

OXIDATIVE C-C CLEAVAGE. Cant and Hall (25) compared the oxidation reactions of ethylene, propylene, 1-butene, cis-2-butene, trans-2-butene, isobutene, and the two 2-pentenes over supported Ir catalysts. The most important oxidation products were acetic acid from ethylene, propene, the 2-butenes, and the 2-pentenes, and propionic acid and acetone from 1-butene and isobutene, respectively. Possible mechanisms are discussed.

COMPLETE COMBUSTION. In a study by Voltz et al. (17) the kinetics of carbon monoxide and propene oxidation on a platinum–alumina catalyst were determined. Complex kinetic equations were formulated, and some rate constants and activation energies were calculated. The oxidation kinetics were used to describe the performance of platinum catalytic converters in automotive emission control systems. The oxidation of propene over Cr(III) and Fe(III) spinels has been investigated by Zanderighi et al. (18). A non-selective oxidation to CO_2 was observed, and a tentative mechanism was advanced. In a study by Yao (19) four types of α-Cr_2O_3 microcrystals were prepared and used as catalysts for the oxidation of C_2H_4, C_3H_6, C_2H_6, C_3H_8, and CO. Reaction rates were measured, and some mechanisms were advanced.

Chemical Kinetic and Physical Aspects. Some of the reviewed papers are discussed according to the following aspects:
proposed mechanisms
nature of active sites
reaction patterns
effects of operating conditions
effect of catalyst modifications and addition of promoters or inhibitors
role of diffusion

PROPOSED MECHANISMS. Kinetics and mechanism of gas–solid catalytic oxidations are generally explained on the basis of redox or Langmuir-Hinshelwood mechanisms or eventually of a third mechanism, which can be considered as combination of the other two. The mechanism according to which the substance to be oxidized reduces the catalyst, which is reoxidized by oxygen from the feed, is known as redox mechanism. This can be assumed to take place in two stages:

$$\text{C-ox} + \text{molecule} \rightarrow \text{C-red} + \text{oxidized molecule} \qquad (1)$$

$$\text{C-red} + O_2 \rightarrow \text{C-ox} \qquad (2)$$

This mechanism has been tested either directly on the basis of data obtained by pulse microreactors (4, 12, 20, 24), in which stage 1 and stage 2 were separated, or by fitting the experimental data from flow reactors to kinetic equations derived from the mechanism (*1, 21, 22, 23*). According to Langmuir-Hinshelwood mechanism the substance to be oxidized and oxygen react together in the adsorbed state. In this case also, either oxidation runs done in pulse microreactors were used to test the proposed mechanism (*12, 18, 26*), or the constants of the kinetic equations based on this mechanism were estimated by the least-squares method (*1, 6, 11*).

Finally the third mechanism, which has been tested in the oxidation of butenes to maleic anhydride (*3, 12, 26*) can be assumed to take place in three stages, which may be written as:

$$\text{C-ox} + \text{butenes} \rightarrow \text{C-red} + \text{butadiene} + H_2O \qquad (1)$$

$$\text{C-red} + O_2 \rightarrow \text{C-red} - O_2 \rightarrow \text{C-ox} \qquad (2)$$

$$\text{C-red} - O_2 + \text{butadiene} \rightarrow \text{maleic anhydride} + CO, CO_2 \qquad (3)$$

Tables II, III, and IV show some reactions for which the various mechanisms were tested.

Table II. Redox Mechanism

Process	Reference
Oxidation of butene to butadiene over various oxide catalysts	*20*
Oxidation of propene to acrolein over Bi–Mo oxide catalysts	*21*
Oxidation of propene to acrylonitrile over U–Sb oxide catalysts	*4*
Oxidation of anthracene to anthraquinone over Co–Mo oxide catalyst	*22*
Oxidation of anthracene to anthraquinone over V oxide catalyst	*23*
Oxidation of methanol to formaldehyde over MoO_3–$Fe_2(MoO_4)_3$ catalyst	*24*
Oxidation of propene to benzene over In oxide catalyst	*1*
Oxidation of butenes to maleic anhydride over $MnMoO_4$ catalyst	*12*

Table III. Langmuir-Hinshelwood Mechanism

Process	Reference
Oxidation of 2-methylpropene to methacrolein over Bi-molybdate catalyst	*6*
Oxidation of ethylene to ethylene oxide over Ag catalyst	*11*
Oxidation of propene to CO_2 over Cr(III) and Fe(III) spinels	*18*
Oxidation of propene to benzene over In oxide catalyst	*1*
Oxidation of butenes to maleic anhydride over $MnMoO_4$ catalyst	*12, 26*

Table IV. Mixed Mechanism

Process	Reference
Oxidation of butenes to maleic anhydride over MnMoO$_4$-based catalyst	12, 26
Oxidation of butene to butadiene over ferrite catalyst	3

NATURE OF ACTIVE SITES. Hypotheses relative to the chemical nature of active sites have been advanced. In the case of ethylene oxidation over Ag catalyst, many authors agree that adsorbed forms of peroxidic molecular oxygen are responsible for the formation of ethylene oxide (27, 28). CO$_2$ formation has been attributed to atomic forms of adsorbed oxygen (27, 28). For the allylic oxidation of olefins and for the selective oxidation of methanol to formaldehyde, lattice oxygen (Me=O type) is proposed as the active oxidizing site (4, 7, 10, 12). The catalytic systems considered are U–Sb oxides (4), molybdates (7, 12), and bismuth tungstates (10). In the oxidation of butenes and butadiene to maleic anhydride over molybdena catalysts, the active sites are assumed to be adsorbed forms of oxygen on Mo(V) or Mo(IV) (12, 13). Finally, for the complete combustion of olefins over various catalysts, adsorbed forms of atomic and molecular oxygen have been suggested (12, 17, 18). In Table V the above mentioned hypotheses are summarized.

REACTION PATTERNS. Except for total oxidation reactions, in all other cases the interesting products are intermediates. It can be expected, therefore, that partial oxidation products are the result of successive or parallel reactions, and from the data in literature there is enough evidence that both mechanisms can be assumed. An example, concerning the oxidation of butenes, is given in Figure 1.

A particular case is the formation of acrylonitrile and maleic anhydride since these products seem to be successive to other stable intermediates—acrolein and butadiene, respectively. This should be true at least in the formation of maleic anhydride, where the conclusion is that

Table V. Nature

Reaction	Catalyst
Oxidation of ethylene to ethylene oxide	Ag
Oxidation of propene to acrylonitrile	U–Sb oxide
Oxidation of 1-butene to butadiene	Fe$_2$O$_3$–MoO$_3$
Oxidation and ammoxidation of olefins	Bi$_2$WO$_6$
Oxidation of butenes to maleic anhydride	MnMoO$_4$
Oxidation of butadiene to maleic anhydride	MoO$_3$
Complete oxidation of propene	Pt
Complete oxidation of propene	Cr(III) and Fe(III) spinels

butadiene is the intermediate (47); however, some controversy still exists for the formation of acrylonitrile. Figure 2 gives a reaction scheme, and some values of calculated kinetic constants are given in Table VI.

EFFECTS OF OPERATING CONDITIONS. From some papers examined in this review it is possible to derive information about the role of operating conditions such as contact time, temperature, partial pressure of oxygen, on yields, selectivities, types of products, etc. The paper by Trifirò et al. (12) on the oxidation of butenes gives a table showing the influence of percent oxygen, temperature, and contact time on the types of reactions occurring. When the amount of oxygen is increased from 0 to 20%, the temperature is increased from 350° to 480°C, and the contact time is increased from 0.27 to 2 sec, the following sequence of catalytic action can be obtained:

isomerization → oxidative dehydrogenation → oxidation with insertion of O_2 → cleavage of C–C bonds and complete oxidation

In an other paper by Trifirò et al. (26) on the same subject the influence of O_2 partial pressure on yields and selectivities is reported.

In a paper by Trimm and Doerr (1) on the oxidation of propene to benzene over indium oxides the yield of major products was examined as a function of contact time and of oxygen and fuel concentrations. They showed that:

The yield of benzene passes through a well-defined maximum at a certain value of the contact time.

The major products—i.e., hexadiene and benzene—increase and pass through a maximum as the concentration of oxygen is increased.

The yield of CO_2 increases linearly with oxygen concentration.

The yield of hexadiene increases with propene concentration while the yield of benzene passes through a maximum and the yield of CO_2 passes through a minimum.

The effect of temperature on the reaction was more complex—the amounts of hexadiene and benzene produced appeared to be inversely

of Active Sites

Reference	Active Sites
27, 28	peroxide molecular oxygen (formation of ethylene oxide) atomic oxygen (formation of CO_2)
4	lattice oxygen
7	lattice oxygen
10	lattice oxygen
12	lattice oxygen
13	oxygen [on Mo(V) or Mo(IV)]
17	atomic and molecular oxygen
18	atomic and molecular oxygen

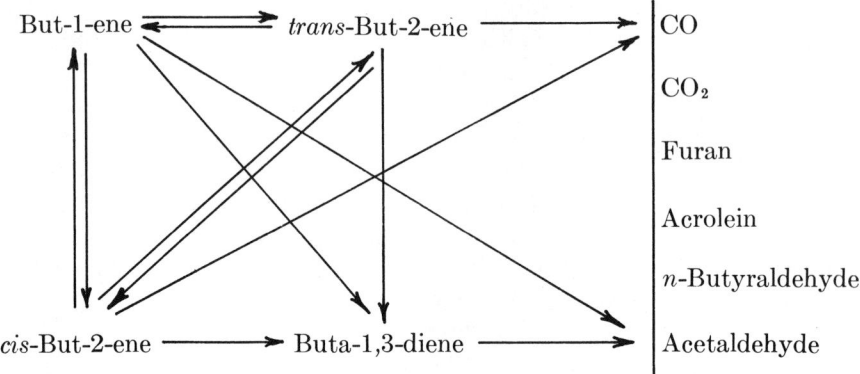

Figure 1. Oxidation of butenes

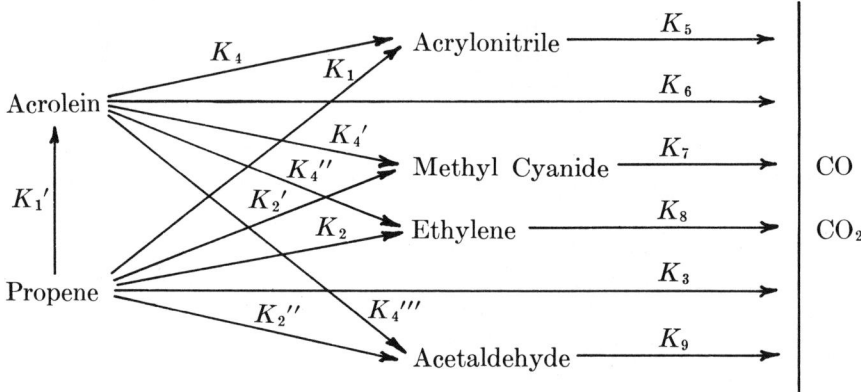

Figure 2. Ammoxidation of propene and acrolein

related. From the above mentioned papers it is possible to deduce that the rate of total oxidation increases as the partial pressure of oxygen is increased.

A different effect of the partial pressure of oxygen was described in a study by Metcalf et al. (11) on silver-catalyzed ethylene oxidation. The rates of both ethylene oxide and carbon dioxide formation passed through maxima with increasing oxygen pressure and then decreased. This behavior was by means of inhibiting effects of the reaction products.

EFFECTS OF CATALYST MODIFICATIONS AND ADDITION OF PROMOTERS OR INHIBITORS. For ethylene oxidation over silver catalysts some authors (27, 28) report that the effect of chlorine is to inhibit dissociative adsorption of oxygen and thus to increase the selectivity of ethylene epoxidation since diatomic oxygen on silver is responsible for the selective oxidation of ethylene to ethylene oxide. The role of Ca and Ba in the same reaction is controversial; some authors (32) contend that the presence of Ca increases selectivity while others (33, 34) believe the addition of Ba and Ca influences only activity.

In the oxidation of 1-butene over Fe_2O_3–MoO_3, Pasquon et al. (7) studied the effect of gaseous oxygen and Te on selectivity. They found that at the lowest oxygen concentrations butadiene and 2-butenes are the main products; when the concentration of oxygen is increased, butadiene and butenes decrease, and maleic anhydride formation reaches a maximum. It was also observed that addition of Te increases the selectivity in butadiene.

ROLE OF DIFFUSION. Three papers (14, 42, 43) deal with effects of diffusion on catalyzed oxidation reactions. For the oxidation of ethylene to acetaldehyde over Pd supported catalysts, Evnin et al. (14) report that the low values determined for the apparent activation energies for ethylene conversion and acetaldehyde formation are an indication of diffusional rather than kinetic control. The presence of diffusional limitations is confirmed by experiments in which the rate was shown to depend on the partial pressure of the nitrogen carrier gas.

In a study on ammonia oxidation over platinum, reported by Pignet and Schmidt (42), a strong influence of mass transfer was observed. Near the stoichiometric composition (21% NH_3 in air) and above, selectivity for NO dropped to zero, in contrast to the reaction in the kinetic regime where significant NO production was observed in excess NH_3 and at temperatures above 1200°C.

Finally, in a study reported by Kadlec et al. (43) the effective diffusivities of air and SO_2 in four industrial vanadium pentoxide catalysts for sulfur dioxide oxidation were measured at the steady state. With models for the effective diffusivity and the kinetics of the catalytic oxidation of SO_2, an optimum apparent density of the catalyst may be determined which gives the maximum rate of reaction per unit volume of catalyst.

Types of Reactors. LABORATORY SCALE. On the laboratory scale pulse microreactors have been used largely to distinguish between redox and Langmuir-Hinshelwood mechanisms. In these studies oxidation runs of organic molecules were done with and without oxygen in the gas phase (4, 7, 12, 18, 20, 24). In the oxidation of butenes to maleic anhydride (7, 12) a Carberry's stirred-tank flow reactor was chosen since heat and mass gradients caused by transfer phenomena are absent in this type of reactor. Conventional type flow reactors were used in most other cases.

INDUSTRIAL SCALE. Literature reports of experimental work in oxidation reactors are scarce, thus making it difficult to verify the mathematical models available. The reactors which dominate in this area are

Table VI. Kinetic

Reaction	K	K_1	K_1'	K_2	K_2'	K_2''	K_3
Propene ammoxidation	5.2	4.8	0.16	0.16	—	0	0.1
Propene oxidation	8.0	—	5.85	—	0	0.06	2.0
Acrolein ammoxidation	170.0	—	—	—	—	—	—
Acrolein oxidation	97.0	—	—	—	0	—	—

the conventional types—*i.e.*, tubular reactors, mutiple tube reactors, multilayer reactors, and fluidized bed reactors. A new type of reactor for gas phase catalytic recycle reactions was recently patented by Collina *et al.* (49). This reactor is sketched in Figure 3 and consists essentially of an injector, followed by a single radial-flux layer of catalyst, and surrounded by an annular empty space through which the reacted gases flow back from the catalyst layer to the injector, exchanging heat with feed gas in a heat exchanger. The conversion per pass is kept very low, and it is therefore easy to control temperature increases. The pressure drop is lower than in conventional reactors, and it is possible to reach high capacities. The reactor is suitable for oxidation reactions such as the oxidation of methanol to formaldehyde, ethylene to ethylene oxide, propene to acrolein and acrylic acid, and butenes to maleic anhydride. Both plant and operating costs for these processes seem competitive with conventional plants.

Papers Presented in This Section. Four of the papers discussed in the last session of the Third International Symposium on Chemical Reaction Engineering (Advances in Chemistry Series No. 133) belong to this section on gas phase catalytic oxidation. Two papers concern the oxidation of *o*-xylene to phthalic anhydride over vanadium catalysts. The other two report studies on the oxidation of carbon monoxide in automotive exhausts.

Oxidation of *o*-Xylene over Vanadium Catalysts. Calderbank (50) studied the kinetics of phthalic anhydride formation in the oxidation of *o*-xylene over a commercial vanadium catalyst. The experimental runs were carried out in a spinning catalyst-basket reactor, and the kinetics were determined for the disappearance of *o*-xylene and of the partial oxidation products *o*-tolualdehyde and phthalide. The kinetics predict observed temperature profiles in large tubular reactors.

Wainwright and Hoffman (51) also studied the kinetics of *o*-xylene oxidation over a vanadium-on-silica catalyst on the basis of experimental data obtained in a laboratory-scale fixed bed reactor and in a pilot-scale transported bed reactor. The kinetic rate expression, which is limited to

Constants

K_4	K_4'	K_4''	K_4'''	K_5	K_6	K_7	K_8	K_9
25.8	—	—	0	0.9	0	1.77	—	—
—	—	—	—	—	6.2	—	—	—
165.0	1	3	0	28	22	—	—	—
—	—	7	42	—	48	—	0	390

o-xylene disappearance, is based on a redox mechanism. The operation of the transported bed reactor has been predicted with good accuracy, and some considerations about selectivities, mechanisms, and physical phenomena, together with suggestions for further research, are included. Agreement between the results reported in these two papers seems to be fairly good. For example the rate expression suggested by Wainwright and Hoffmann contains a parameter θ, which becomes equal to 1 with a fully oxidized catalyst, in agreement with the disappearance kinetics of o-xylene proposed by Calderbank, which are first order up to nearly 1 mole % of xylene concentration. The observations regarding the activation energy are also similar in both papers. The activation energy decreased as the reaction temperature increased, becoming about a third of that at lower temperatures. The operation temperature range, however, is considerably lower in Wainwright and Hoffman's study, possibly because of the different catalysts used.

OXIDATION OF CARBON MONOXIDE IN AUTOMOTIVE EXHAUSTS. The catalytic oxidation of carbon monoxide has been studied extensively with many catalysts. Noble catalysts have received considerable attention during the last several years for use in automotive emission control systems. A kinetic model of CO and C_3H_6 oxidation on a pelleted platinum–alumina catalyst was prepared and incorporated into a previously developed converter model, which has been used to predict and optimize the performance of various types of platinum catalysts (including monolithic types); this mathematical converter model was described by Kuo et al. (52, 53).

The oxidation of CO over platinum catalysts has been studied for many years, but the conclusions regarding the mechanisms and rate equations are somewhat conflicting (54, 55, 56, 57). Shishu (58) made a comprehensive investigation of CO oxidation on a monolithic platinum catalyst in a differential flow reactor and developed rate equations. Harned (59) prepared a mathematical model for catalytic converters used for the oxidation of CO and hydrocarbons, and Taylor (60) obtained a model of the oxidation kinetics for platinum catalysts. CO oxidation has been studied with other catalysts also. Laidler (61) reported some studies done with catalysts such as quartz glass, rock crystal, platinum, and copper oxide. Schwab and Gossner (62) used silver, palladium, and silver–palladium alloys. Parravano (63) reported the use of nickel oxide, to which foreign ions had been added. α-Alumina pellets with 0.5 wt % palladium and x-ray irradiated alumina have also been proposed as catalysts (64, 65).

At present, exhaust treatment devices, such as catalytic converters show promise in reducing CO emissions to the low levels required by the restrictions imposed in many countries. The catalysts to be used by auto manufacturers in the next years generally contain platinum, which is, however, expensive and easily poisoned. Thus, there is a need for cheaper catalysts with good performance and of reliable catalytic converter models. The papers dicussed below deal with these two problems.

Kalman et al. (66) evaluated the performance of crystalline copper-substituted zirconium phosphate in the catalytic oxidation of CO. A reaction rate expression was obtained by applying the integral method of analysis to the experimental data. Even though a direct comparison of the performance of this catalyst with other catalysts is difficult owing to the lack of literature data obtained under similar conditions, it seems that copper-substituted α-zirconium phosphate is at least comparable in activity with other catalysts and merits further investigation .

Young and Finlayson (67) propose two mathematical models for a particular type of catalytic converter—the monolith converter. Two types of catalytic converters—packed beds and monoliths—have been proposed for the oxidation of CO and hydrocarbons in automobile exhausts. As mentioned above, Kuo has developed a model for the monolith converter, and mathematical models have been proposed and solved also for packed bed converters (52, 59, 68, 69). The monolith converter consists of an array of ducts or cells through which the exhaust gas flows axially. Because of a smaller volumetric heat capacity, monolithic converters warm up more quickly than packed bed devices, but have problems caused by thermal expansion. The mathematical models developed illustrate important features of the thermal behavior of the monolith converter. The reaction rate expression used was found in the literature, and some assumptions were made to obtain suitable models, taking into account heat and mass transfer across the cell cross-section. The two proposed models differ because in one of them heat and mass transfer coefficients defined in the usual way are used. Suitable solution methods have been used for both models, and both transient and steady-state calculations for typical cases are illustrated in some figures. The study seems to be very accurate and gives useful advice for designing this type of converter.

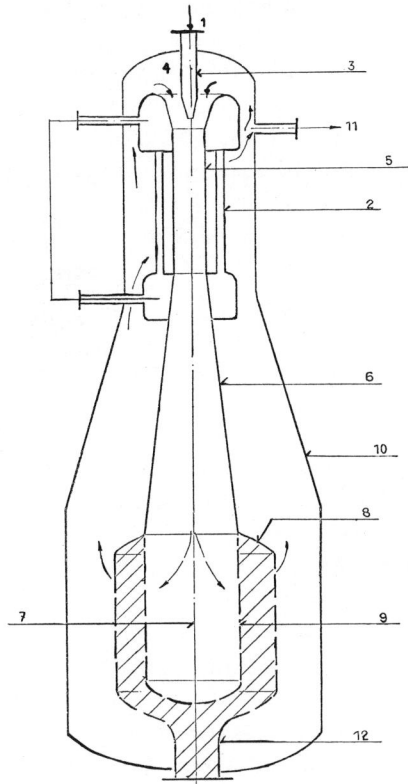

Figure 3. Reactor for catalytic gas phase reactions

Liquid Phase Oxidation with Oxygen

In this group of processes both homogeneous and heterogeneous catalyzed reactions are presented. The liquid-phase oxidation of organic compounds with air or oxygen is complex, and the mechanisms are further complicated by mass transfer processes. When oxygen transfer becomes the rate-limiting step, the rate of the overall process is no longer controlled by the chemical mechanisms.

A paper by Hobbs *et al.* on mass transfer rate-limitation effects in liquid phase oxidation (45) indicates that physical and chemical effects are theoretically separable, and their relative contributions can be estimated. As an example, the liquid phase oxidation of methyl ethyl ketone in acetic acid solvent is considered.

In the study reported by Gurumurthy and Govindarao (46) a rate equation is developed for the liquid phase oxidation of propionaldehyde with oxygen in the presence of manganese propionate catalyst. de Wilt and Van der Baan (44) present a kinetic model for the platinum catalyzed oxidation of glucose to k-gluconate with oxygen in aqueous alkaline solutions. The experiments were done batchwise in a reactor equipped with a high-speed stirrer to minimize the influence of oxygen transport from the gaseous to the liquid phase. The reactions studied are listed in Table VII.

Table VII. Liquid Phase Oxidation with Oxygen

Process	Catalyst	Reference
Oxidation of propionaldehyde	Mn–propionate	46
Oxidation of glucose	platinum	44
Oxidation of methyl ethyl ketone	cobaltous acetate	45

Alagy et al. (70) describe a study on the liquid phase oxidation of cyclohexane with oxygen. A mathematical model has been developed on the basis of kinetic data concerning cyclohexane oxidation and mass-transfer information derived from experiments on cyclododecane oxidation. The experimental runs were done in semicontinuous, mechanically stirred equipment. On the pilot scale a 150-liter cylindrical reactor was used, in which stirring was achieved by injecting gas at the bottom of the column and by introducing tangentially at the top a liquid stream derived from the bottom of the column. The kinetic model has seven parameters which have been determined by a non-linear regression method. The agreement between calculated and experimental data seems good, taking into account that the reaction is complex and both chemical and physical phenomena are important. A simplified model has also been prepared by incorporating certain assumptions in the original model.

Liquid Phase Epoxidation of Olefins with Hydroperoxides

Four papers (35, 36, 37, 38) report studies on liquid phase epoxidation of olefins with hydroperoxides. The reaction is:

$$\diagdown\!\!\!\!\diagup C = C \diagup\!\!\!\!\diagdown + ROOH \longrightarrow \diagdown\!\!\!\!\diagup C \underset{O}{-} C \diagup\!\!\!\!\diagdown + ROH$$

and is catalyzed by many transitional metal catalysts. The kinetic equation, which has been obtained for different molybdenum catalysts is of the type:

$$v = K \, C_{\text{olefin}} \, C_{\text{hydroperoxide}} \, C_{\text{catalysts}}$$

and is catalyzed by many transition metal catalysts. The kinetic equa-

equilibrium formation of a catalyst–hydroperoxide complex; the second step is the rate-determining reaction of the complex with the olefin to form the epoxide, coproduct alcohol, and the molybdenum catalyst.

Baker et al. (35) present kinetic data on the epoxidation of 1-octene by cumene and *tert*-butyl hydroperoxide in the presence of molybdenum hexacarbonyl as catalyst. The observed kinetic behavior is compatible with the previously proposed general mechanism for the epoxidation reaction.

In the study reported by Sheldon and Van Doorn (36) cyclohexene and 1-octene, used as model olefins of different reactivity, were epoxidized in the liquid phase with *tert*-butyl hydroperoxide in the presence of various transition metal catalysts. It is concluded that an active epoxidation catalyst should be both a weak oxidant and a fairly strong Lewis acid. These requirements are best met by compounds of certain metals in high oxidation states [Mo(VI), W(VI), Ti(IV)].

In the paper by Su et al. (37) on vanadium and molybdenum chelates as catalysts in the epoxidation of cycloalkenes, rate laws for the vanadium-catalyzed systems are consistent with reaction *via* rate-determining attack of olefin on a vanadium(V)–hydroperoxide complex. Arguments are presented to support the view that the molybdenum-catalyzed epoxidation, like those involving vanadium, proceeds by reaction of olefins with a metal–hydroperoxide complex.

In a study by Trifirò et al. (38) on the liquid phase epoxidation of cyclohexene by *tert*-butyl hydroperoxide on a Mo-based catalyst, a rate law is given, and the presence of a catalyst–hydroperoxide reversible complex as the active species in the epoxidation is advanced on the basis of a spectroscopic study. The reactions investigated are listed in Table VIII.

Heterogeneization of Homogenous Catalytic Processes

Some studies have been reported on the transformation of homogeneous processes into heterogeneous ones by supporting the catalysts. This is the case of a heterogeneous catalyst system which has been developed for the vapor phase oxidation of ethylene to acetaldehyde (14). The catalyst consists of palladium-doped vanadium pentoxide and usually a third component such as Ti, Ru, Pt, or Ir. A catalyst consisting of mercuric chloride supported on active charcoal was used by Arai et al. (16) for the allylic oxidation of olefins. The oxidation over mercuric ion catalyst occurs at a temperature which is lower than that necessary over bismuth molybdate catalyst. Finally, catalysts prepared by supporting Mo(VI) on SiO_2 have been proposed by Forzatti et al. (40) for the epoxidation of olefins with hydroperoxides.

Evaluation of the Literature Data

This section focuses on those aspects of interest to chemical reaction engineers in modeling reactions and in designing industrial reactors. The foregoing will have given the reader an appreciation not only of what has been done but of the criteria which are generally followed by the authors active in this area. The points examined under chemical kinetic and physical aspects, for example, are typical of these kinds of papers and account for the fact that the interests of the authors are fundamentally chemical. Most papers deal with microscopic mechanisms, and several specific chemical systems are used to illustrate these mechanisms, but generalized reaction models are seldom given. Often, detailed and elegant studies concerning the catalytic chemistry of the reactions are carried out but useful models for reactor design are not attempted. The range of catalysts considered is extensive, but some differ in irrelevant details.

Discussion of the details of surface bonds can be found in many papers, and parametric effects of operating conditions on activity and selectivity are often discussed, but mostly on the laboratory scale. On the other hand, it is not easy to find kinetic studies done to provide reliable data to predict reactor performance, and systematic techniques in chemical kinetics investigation are not often applied. The use of statistically designed experiments, for example, is mentioned in only a

Table VIII. Liquid Phase Epoxidation with Hydroperoxides

Process	*Catalyst*	*Reference*
Epoxidation of 1-octene by cumene and *tert*-butyl hydroperoxide	Mo–hexacarbonile	*35*
Epoxidation of 1-octene and cyclohexene by *tert*-butyl hydroperoxide	transition metals	*36*
Epoxidation of cycloalkenes by *tert*-butyl hydroperoxide	V and Mo chelates	*37*
Epoxidation of cyclohexene by *tert*-butyl hydroperoxide	Mo based	*38*

few papers, and it does not seem that advanced parameter estimation techniques are often used also if a certain improvement from this point of view can be noticed. There is lack of information concerning studies carried out sequentially on the laboratory, pilot, and commercial scale. In this connection most of the papers come from universities rather than from industries, and the possibility of obtaining valuable information on process kinetics from commercial scale reactors is generally disregarded, at least for publication purposes. Nevertheless, some authors seem to be more and more conscious of the necessity of providing useful data for design purposes; in those fields which are more strongly connected with solving practical problems (as, for example, in automative emission control), reliable models are available.

Conclusions

The situation in this area has some positive aspects but is not entirely satisfactory from the point of view of a chemical engineer. The quality of papers found in literature is generally good, but most studies are based on laboratory scale experiments, and scale-up rules are seldom given. In this area some authors still "play with mechanisms" and carry out experiments with different catalysts and under different operating conditions without arriving at conclusions of any sort. However, a definite improvement has been evidenced in the last few years; in most cases now at least kinetic models, useful from an engineering point of view, are proposed as a result of a research study. Nevertheless, the communication between those who propose models and those who should use them does not seem entirely satisfactory. The bridge between researchers and those involved in design and operation of reactors, that Professor Froment mentioned at the end of a review during the first International Symposium on Chemical Reaction Engineering, is still a weak bridge of boats, and not even the observation that the Romans were able to unify the western world using this type of bridge can reassure us entirely in the 1970's.

Literature Cited

1. Trimm, D. L., Doerr, L. A., *J. Catalysis* (1972) **26**, 1.
2. Parera, N. S., Trimm, D. L., *J. Catalysis* (1973) **30**, 485.
3. Sterrett, J. S., Hollvried, H. G., *Ind. Eng. Chem., Process Design Develop.* (1974) **13**, 54.
4. Grasselli, R. K., Suresh, D. D., *J. Catalysis* (1972) **25**, 273.
5. Wragg, R. D., Ashmore, P. G., Hockey, J. A., *J. Catalysis* (1973) **31**, 293.
6. Mann, R. S., Ko, D. W., *J. Catalysis* (1973) **30**, 276.
7. Pasquon, I., Trifirò, F., Caputo, G., *Chim. Ind.* (1973) **55**, 168.
8. Schuit, G. C. A., paper presented at the Conference on The Chemistry and Uses of Molybdenum, University of Reading, England, Sept. 1973.
9. Daniel, C., Keulks, G. W., *J. Catalysis* (1973) **29**, 475.
10. Villa, P. L., Caputo, G., Sala, F., Trifirò, F., *J. Catalysis* (1973) **31**, 200.
11. Metcalf, P. L., Harriot, P., *Ind. Eng. Chem., Process Design Develop.* (1972) **11**, 478.
12. Trifirò, F., Banfi, C., Caputo, G., Forzatti, P., Pasquon, I., *J. Catalysis* (1973) **30**, 393.
13. Akimoto, M., Echigoya, E., *J. Catalysis* (1973) **29**, 191.
14. Evnin, A. B., Rabo, J. A., Kasai, P. H., *J. Catalysis* (1973) **30**, 109.
15. Giordano, N., Vaghi, A., Bart, J. C. J., Castellani, A., Symposium on the Mechanisms of Hydrocarbon Reactions, Siojok, June 1973.
16. Arai, H., Uehara, K., Kinoshita, S., Kunugi, T., *Ind. Eng. Chem., Prod. Res. Develop.* (1972) **11**, 308.
17. Voltz, S. E., Morgan, C. R., Liederman, D., Jacob, S. M., *Ind. Eng. Chem., Prod. Res. Develop.* (1973) **12**, 294.

18. Zanderighi, L., Faedda, M. P., Carrá, S., *J. Catalysis*, in press.
19. Yao, Y. Y., *J. Catalysis* (1973) **28**, 39.
20. Niwa, M., Murakami, Y., *J. Catalysis* (1972) **27**, 26.
21. Gernan, K., Grzybowska, B., Haber, J., *Bull. Acad. Polonaise Sci.* (1973) **21**, 5.
22. Subramanian, P., Murthy, M. S., *Chem. Eng. Sci.* (1974) **29**, 25.
23. Subramanian, P., Murthy, M. S., *Ind. Eng. Chem., Process Design Develop.* (1972) **11**, 242.
24. Liberti, L., Pernicone, N., Soattini, S., *J. Catalysis* (1972) **27**, 52.
25. Cant, N. W., Hall, W. K., *J. Catalysis* (1972) **27**, 70.
26. Trifirò, F., Caputo, G., Villa, P. L., *J. Less Common Metals* (1974) **36**, 305.
27. Kilty, P. A., Rol, N. C., Sachtler, W. M. H., *Proc. Intern. Congr. Catalysis, 5th, Miami Beach*, V. S. D. Elsevier (1972) **64**, 929.
28. Spath, H. T., *Proc. Congr. Catalysis, 5th, Miami Beach*, V. S. D. Elsevier (1972) **65**, 945.
29. Marcinkowsky, A. E., Berty, J. M., *J. Catalysis* (1973) **29**, 494.
30. Carberry, J. J., Kuczynski, G. C., Martinez, E., *J. Catalysis* (1972) **26**, 246.
31. Forzatti, P., Martinez, E., Kuczynski, G. C., Carberry, J .J., *J. Catalysis* (1973) **28**, 455.
32. Forzatti, P., Klimasara, A., Kuczynski, G. G., Carberry, J. J., *J. Catalysis* (1973) **19**, 169.
33. Spath, H. T., Tomazic, G. S., Würm, H., Torkar, K., *J. Catalysis* (1972) **26**, 18.
34. Spath, H. T., Torkar, K., *J. Catalysis* (1972) **26**, 163.
35. Baker, T. N., Mains, G. J., Sheng, M. N., Zatacek, J. G., *J. Org. Chem.* (1973) **38**, 1145.
36. Sheldon, R. A., Van Doorn, J. A., *J. Catalysis* (1973) **31**, 427.
37. Su, C. C., Reed, J. W., Gould, E. S., *Inorg. Chem.* (1973) **12**, 337.
38. Trifirò, F., Forzatti, P., Preite, S., Pasquon, I., *J. Less Common Metals* (1974) **36**, 319.
39. Fujimoto, K., Negami, Y., Takahashi, T., Kunugi, T., *Ind. Eng. Chem., Prod. Res. Develop.* (1972) **11**, 303.
40. Forzatti, P., Trifirò, F., Pasquon, I., *Chim. Ind.* (1974) **56**, 259.
41. Walsh, M. A., Katzer, J. R., *Ind. Eng. Chem., Process Design Develop.* (1973) **12**, 477.
42. Pignet, T., Schmidt, L. D., *Chem. Eng. Sci.* (1974) **29**, 1123.
43. Kadler, B., Hudgins, R. R., Silveston, P. L., *Chem. Eng. Sci.* (1973) **28**, 935.
44. De Wilt, H. G. J., Van Der Baan, H. S., *Ind. Eng. Chem., Prod. Res. Develop.* (1972) **11**, 374.
45. Hobbs, C. C., Drew, E. H., Van't Hoff, H. A., Mesich, F. G., Onorem, J., *Ind. Eng. Chem., Prod. Res. Develop.* (1972) **11**, 220.
46. Gurumurthy, C. C., Govindarao, V. M. H., *Ind. Eng. Chem., Fundamentals* (1974) **13**, 9.
47. Dente, M., Ranzi, E., Quiroga, S. P., Biardi, G., *Chim. Ind.* (1973) **55**, 563.
48. Pitzer, E. W., *Ind. Eng. Chem., Prod. Res. Develop.* (1972) **11**, 299.
49. Collina, A., Malfatti, E., Cappelli, A., Italian Patent **955,507** corresponding to German Patent Application **2,324,164**.
50. Calderbank, P. H., ADVAN. CHEM. SER. (1974) **133**, 646.
51. Wainwright, M. S., Hoffmann, T. W., ADVAN. CHEM. SER. (1974) **133**, 669.
52. Kuo, J. C. W., Horgan, C. R., Lassen, H. G., *SAE Automotive Eng. Congr., Detroit, Jan. 1972*, paper 710289.
53. Kuo, J. C. W., Prater, C. D., Osterhout, D. P., Snyder, P. W., Wej, J., *Int. Automobile Tech. Congr. FISITA, 14th, London, June 1972*, paper 2/14.
54. Langmuir,.I., *Trans. Faraday Soc.* (1922) **17**, 621.
55. Sklyarov, A. V., Tret'yakov, I. I., Shad, B. R., Roginski, S. Z., *Dokl. Phys. Chem.* (1969) **189**, 829.

56. Su, E. C., Shishu, R. C., private communication, 1972.
57. Soloveva, L. S., *Russian J. Phys. Chem.* (1960) **34**, 586.
58. Shishu, R. C., Ph.D. Dissertation, University of Detroit (1972).
59. Harned, J. L., *SAE Nat. Automotive Eng. Congr., Detroit, May 1972*, paper 720520.
60. Taylor, K. C., unpublished results (1971).
61. Laidler, J. K., *Catalysis* (1954) **1**, 16.
62. Schwab, G. M., Gossner, I. A., *Phys. Chem., N.F.* (1958) **16**, 39.
63. Parravano, G., *J. Am. Chem. Soc.* (1953) **75**, 1448.
64. Taibl, D. B., Simons, J. B., Carberry, J. J., *Ind. Eng. Chem., Fundamentals* (1966) **5**, 171.
65. Coekelbergs, R., Collin, R., Cruez, A., Decot, J., Degols, L., Timmerman, L., *J. Catalysis* (1967) **7**, 85.
66. Kalman, T. J., Duovkovic, M., Clearfield, A., ADVAN. CHEM. SER. (1974) **133**, 654.
67. Young, L. C., Finlayson, B. A., ADVAN. CHEM. SER. (1974) **133**, 629.
68. Ferguson, N. B., Finlayson, D. A., *AIChE Meetg., Philadelphia, Nov. 1973*.
69. Wei, J., *Chem. Eng. Progr., Monograph Ser.* (1969) **6**, 65.
70. Alagy, J., Trambouze, P., Van Landeghem, H., ADVAN. CHEM. SER. (1974) **133**, 644.

RECEIVED December 4, 1974.

INDEX

A

Acetylene, hydrogenation of	52
Acrolein, ammoxidation of	220
Adsorption isotherms	29
Air, vapor phase oxidation with	212
Allylic oxidation	213
Ammonia oxidation over platinum	221
Ammonia synthesis	34
kinetics of	122
Ammoxidation of propene and acrolein	220
Applications, biological	199
Arrhenius kinetics	159
Autoclave, stirred	59
Automotive exhausts, oxidation of carbon monoxide in	223
Axial conduction	189
Axial dispersion	68, 88
Axial mixing	187

B

Battelle agglomerating bed gasifier, Union Carbide/	149
Beds, dispersion effects in packed	80
Beds, fluidized	88
Beds, gas-liquid flow in packed	126
Behavior, hydrodynamic	60, 65
Belousov-Zhabotinskii reaction	200
Benzene, hydrogenation of	54, 118
Bi-Gas reactor	151
Bimolecular Langmuir kinetics	16
Biological applications	199
Boundary conditions for the fixed bed tubular reactor	93
Boundary layer problem	198
Bubbles	89
Bureau of Mines Sythane coal gasification	150
Butane, oxidation of	20
1-Butene	213, 215
oxidation of	221
Butynediol	52

C

CSTR	18, 158
California Cycle	21
Carbon conversion	143
Carbon monoxide in automotive exhausts, oxidation of	223
Cascade model, heterogeneous	9
Cascade model, pseudo-homogeneous	8
Catalysis, kinetic models in heterogeneous	26
Catalyst	110
decay	110
durability	22
effectiveness factor	70
particle problem	171, 173
particles, single	183
phase	186
poisoning	60
screens	90
vanadium-on-silica	222
Catalysts	
molybdate	213
porous	173
Catalytic	
beds	6
monolithic	10
converters	224
cracking models	107
gas phase reactions, reactors for	225
muffler	1
processes, heterogeneization of homogeneous	227
reforming	101
wires and gauzes	182
Char	135
—steam reaction	149
Chemical processing	52
Chromatography, gas	88
Clean Air Act	1, 5
Coal gasification	132
Coal–hydrogen reactions	140
CO oxidation over platinum and palladium	15
CO_2 acceptor process	148
Combustion	201
complete	216
Commercial hydrane reactor	146
Compleat process model	99
Complete combustion	216
Conditions for uniqueness	187
Conduction, axial	189
Contacting effectiveness	60
in trickle bed reactors	50
Control, stabilizing	164
Conversion, carbon	143
Conversion, para-ortho hydrogen	36
Cycloalkenes, epoxidation of	227
Cyclohexene	227
Cyclopropane, hydrogenation of	37
Cyclopropane, isomerization of	54
Cracking models, catalytic	107
Crotonaldehyde	71
hydrogenation of	54

D

Decay, catalyst	110
Dehydration of ethanol to ether	36
Dehydrodimerization	213
Dehydrogenation, oxidative	213
Design, reactor	153
Destructive melting of monoliths	21
Diffusion	221
and dispersion, packed bed	81
equations, dimensionless	171
-limited reaction	71
systems	77
Dimensionless diffusion equations	171
Dispersion	
axial	68, 88
effects in packed beds	80
with mass transfer between phases	88
packed bed	81
Taylor	82
Dynamics	20
tubular and fixed-bed reactor	194

E

Effectiveness, contacting	60
Effectiveness factor, catalyst	70
Electrochemical systems	202
Efficiency, reactor	67
Engineering, physical processes in chemical reactor	75
Epoxidation of cycloalkenes	227
Epoxidation of olefins with hyperoxides	226
Equations, dimensionless diffusion	171
Ethanol to ether, dehydration of	36
Ethylene oxidation	220
Exothermic reaction, single	158

F

Feedback, heat	196
Fischer-Tropsch synthesis	52
Fixed-bed reactors, tubular and	185, 194
Fixed-bed tubular reactor, boundary conditions for	93
Flow in packed beds, gas-liquid	126
Flow model, pulsing	128
Flow systems	76
Fluid bed models	120
Fluidized beds	88
Freundlich and Temkin isotherms	33

G

Gas	
chromatography	88
generator, Wellman-Galusha fuel	144
low-Btu	138
process, high-Btu	141
producer	138
purification	137
SNG and synthesis	134
Gasification, Bureau of Mines Synthane coal	150
Gasification, coal	132
Gasification reaction rates	139
Gasifier, Koppers-Totzek	145
Gasifier, Lurgi pressure	142
Gasifier, Union Carbide/Battelle agglomerating bed	149
Gauzes, catalytic wires and	182

H

Heat and mass transfer	117
Heat feedback	196
Heat transfer effects	70
Heterogeneization of homogeneous catalytic processes	227
Heterogeneous cascade model	9
Heterogeneous catalysis, kinetic models in	26
Heterogeneous models	189
Heterogeneous reaction systems	76
High-Btu gas processes	141
Holdup	66, 67
liquid	62
Homogeneous catalytic processes, heterogeneization of	227
Hydrocracking	51, 54, 63
process model	104
Hydrane reactor, commercial	146
Hydrodenitrogenation	54, 63
Hydrodesulfurization	51, 57, 62, 63
Hydrodynamic behavior	60, 65
Hydrogasification reactor section, IGT	147
Hydrogen peroxide manufacturing	52
Hydrogen reactions, coal–	140
Hydrogenation	54
of acetylene	52
of benzene	54, 118
of crotonaldehyde	54
of cyclopropane	37
of α-methylstyrene	54
of propylene and isobutylene	39
Hydrogenolysis reactions	112
Hydroperoxides, liquid phase epoxidation of olefins with	226
Hydrotreating	51, 54

I

IGT hydrogasification reactor section	147
Ideal trickle bed reactor	58
Industrial operating conditions	54
Industrial process models	98
Industry, petroleum	54
Instabilities	183
Isobutene	215
Isobutylene, hydrogenation of propylene and	39
Isomerization of cyclopropane	54
Isotherm, adsorption	29
Isotherm, Langmuir	30
Isotherms, Freundlich and Temkin	33

INDEX

Isothermal liquid-phase Belousov-
  Zhabotinskii reaction 201
Isothermal systems 164

K

Kinetic models 28
  in heterogeneous catalysis 26
Kinetics
  ammonia synthesis 122
  Arrhenius 159
  bimolecular Langmuir 16
  lumped 101
  negative order 15
Kmak's process model 103

L

Langmuir-Hinshelwood mechanism 217
Langmuir isotherm 30
Langmuir kinetics, bimolecular ... 16
Liquid
  flow in packed beds, gas– 126
  holdup 62
  hourly space velocity (LHSV) . 56
  phase epoxidation of olefins with
    hydroperoxides 226
  phase oxidation with oxygen ... 225
Low-Btu gas 138
Lumped kinetics 101
Lumped models 178
Lumping, methods of 178
Lumping, theoretical analysis of .. 113
Lurgi pressure gasifier 142

M

Manganese propionate 226
Mars-van Krevelen models 44
Mass transfer between phases,
  Taylor dispersion with 88
Mass transfer effects 70
Mass transfer, heat and 117
Methane 136
  generation, rate of 147
Methods of lumping 178
α-Methylstrene 71
  hydrogenation of 54
Mixing 196
  axial 187
Model
  building, statistical 38
  compleat process 99
  Kmak's process 103
  hydrocracking process 104
Modeling 196
Models
  catalytic cracking 107
  fluid bed 120
  heterogeneous 189
  in heterogeneous catalysis,
    kinetic 26
  industrial process 98
  kinetic 28
  lumped 178

Molybdate catalysts 213
Molybdenum chelates, vanadium
  and 227
Monolithic catalytic beds 10
Monoliths, destructive melting of . 21
Moving bed reactor 109
Muffler, catalytic 1
Multiplicity 196
  and stability in the CSTR,
    steady-state 166
  stability, and sensitivity of states
    in chemical reacting systems 156
  steady-state 183

N

Naphthas, petroleum 101
Negative order kinetics 15

O

1-Octene 227
Olefins with hydroperoxides, liquid
  phase epoxidation of 226
Optimum sloppines principle 99
Oscillatory states 183
Oxidation
  allylic 213
  of butane 20
  of 1-butene 221
  of carbon monoxide in automo-
    tive exhausts 223
  ethylene 220
  with oxygen or air, vapor phase 212
  with oxygen, liquid phase 225
  over platinum, ammonia 221
  of propene 219
  reaction engineering 212
  of SO_2 ....54, 79
  of o-xylene 42
    over vanadium catalysts 222
Oxidative C–C cleavage 216
Oxidative dehydrogenation 213
Oxygen insertion 215
Oxygen, liquid phase oxidation
  with 225
Oxygen or air, vapor phase oxidation with 212

P

Packed bed diffusion and disper-
  sion 80, 81
Packed beds, gas–liquid flow in ... 126
Packed beds, shallow 7
Palladium 3
  CO oxidation over platinum and 15
Para-ortho hydrogen conversion .. 36
Particle problem, catalyst 173
Peclet number 69, 87
Petrochemical processing 52
Petroleum industry 54
Petroleum naphthas 101
Petroleum refining 50, 56
Phase, catalyst 186

Phase diagrams 162
Physical processes in chemical reactor engineering 75
Platinum 3
   ammonia oxidation over 221
   and palladium, CO oxidation over 15
Poisoning, catalyst 60, 120
Porous catalysts 173
Power rate laws 41, 46
Process, high-Btu gas 141
Process model, the compleat 99
Producer gas 138
Propene, ammoxidation of 220
Propene, oxidation of 219
Propylene and isobutylene, hydrogenation of 39
Psuedo-homogeneous cascade model 8
Pulsing flow model 128
Purification, gas 137

Q
Quinoline 55

R
Rate equations 34
Rate laws, power 41
Rate of methane generation 147
Rates, gasification reaction 139
Reaction
   Belousov-Zhabotinskii 200, 201
   char–steam 149
   diffusion-limited 71
   engineering, oxidation 212
   rates, gasification 139
   single exothermic 158
   systems, heterogeneous 76
Reactions, coal–hydrogen 140
Reactions, hydrogenolysis 112
Reactor
   Bi-gas 151
   boundary conditions for fixed bed tubular 93
   for catalytic gas phase reactions 225
   commercial hydrane 146
   design 153
   dynamics, tubular and fixed-bed 194
   efficiency 67
   engineering, physical processes in chemical 75
   ideal trickle bed 58
   moving bed 109
Reactors
   contacting effectiveness in trickle bed 50
   recycle 18
   slurry 53
   transport 122
   tubular and fixed bed 185
   types of 221
Redox mechanism 217

Refining, petroleum 50, 56
Reforming, catalytic 101
Residence time distribution 66

S
SNG and synthesis gas 134
SO_2 oxidation 54, 79
Screens and shallow beds, catalyst 90
Sensitivity 156, 192
Shallow packed beds 7
Single catalyst particles 183
Single exothermic reaction 158
Slurry reactors 53
Stability 156, 196
   in the CSTR 166
   of steady-state solutions 175
Stabilizing control 164
Statistical model building 38
Steady-state multiplicity 183
   and stability in the CSTR 166
Steady-state solutions, stability of . 175
Steady-state, uniqueness of 161
Steam reaction, char– 149
Sythane coal gasification, Bureau of Mines 150
Synthesis, Fischer-Tropsch 52
Synthesis kinetics, ammonia 122

T
Taylor dispersion 82
   with mass transfer between phases 88
Temkin isotherms, Freundlich and 33
Theoretical analysis of lumping .. 113
Tolualdehyde 42
Transience 20
Transport reactors 122
Trickle bed reactor, ideal 58
Trickle bed reactors, contacting effectiveness in 50
Tubular and fixed-bed reactors 93, 185, 194
Types of reactors 221

U
Union Carbide/Battelle agglomerating bed gasifier 148
Uniqueness, conditions for 187
Uniqueness of a steady state 161

V
Vapor phase oxidation with oxygen or air 212
Vanadium and molybdenum chelates 227
Vanadium catalysts 222

Vanadium pentoxide	221	Wetting characteristics	64
Vanadium-on-silica catalyst	222	Wires and gauzes, catalytic	182

W

Wellman-Galusha fuel gas generator 144

X

o-Xylene, oxidation of 42
    over vanadium catalysts 222

The text of this book is set in 10 point Caledonia with two points of leading. The chapter numerals are set in 30 point Garamond; the chapter titles are set in 18 point Garamond Bold.

The book is printed offset on Text White Opaque, 50-pound. The cover is Joanna Book Binding blue linen.

*Jacket design by Norman Favin.
Editing and production by Mary Rakow.*

The book was composed by the Mills-Frizell-Evans Co. and by Service Composition Co., Baltimore, Md., printed and bound by The Maple Press Co., York, Pa.